Side lights on Astronomy and Kindred Fields of Popular Science

SIMON NEWCOMB

Published in 1907

TABLE OF CONTENTS

PREFACE

In preparing and issuing this collection of essays and addresses, the author has yielded to what he could not but regard as the too flattering judgment of the publishers. Having done this, it became incumbent to do what he could to justify their good opinion by revising the material and bringing it up to date. Interest rather than unity of thought has determined the selection.

A prominent theme in the collection is that of the structure, extent, and duration of the universe. Here some repetition of ideas was found unavoidable, in a case where what is substantially a single theme has been treated in the various forms which it assumed in the light of constantly growing knowledge. If the critical reader finds this a defect, the author can plead in extenuation only the difficulty of avoiding it under the circumstances. Although mainly astronomical, a number of discussions relating to general scientific subjects have been included.

Acknowledgment is due to the proprietors of the various periodicals from the pages of which most of the essays have been taken. Besides Harper's Magazine and the North American Review, these include McClure's Magazine, from which were taken the articles "The Unsolved Problems of Astronomy" and "How the Planets are Weighed." "The Structure of the Universe" appeared in the International Monthly, now the International Quarterly; "The Outlook for the Flying-Machine" is mainly from The New York Independent, but in part from McClure's Magazine; "The World's Debt to Astronomy" is from The Chautauquan; and "An Astronomical Friendship" from the Atlantic Monthly.

SIMON NEWCOMB. WASHINGTON, JUNE, 1906.

THE UNSOLVED PROBLEMS OF ASTRONOMY

The reader already knows what the solar system is: an immense central body, the sun, with a number of planets revolving round it at various distances. On one of these planets we dwell. Vast, indeed, are the distances of the planets when measured by our terrestrial standards. A cannon-ball fired from the earth to celebrate the signing of the Declaration of Independence, and continuing its course ever since with a velocity of eighteen hundred feet per second, would not yet be half-way to the orbit of Neptune, the outer planet. And yet the thousands of stars which stud the heavens are at distances so much greater than that of Neptune that our solar system is like a little colony, separated from the rest of the universe by an ocean of void space almost immeasurable in extent. The orbit of the earth round the sun is of such size that a railway train running sixty miles an hour, with never a stop, would take about three hundred and fifty years to cross it. Represent this orbit by a lady's finger-ring. Then the nearest fixed star will be about a mile and a half away; the next more than two miles; a few more from three to twenty miles; the great body at scores or hundreds of miles. Imagine the stars thus scattered from the Atlantic to the Mississippi, and keep this little finger-ring in mind as the orbit of the earth, and one may have some idea of the extent of the universe.

One of the most beautiful stars in the heavens, and one that can be seen most of the year, is a Lyrae, or Alpha of the Lyre, known also as Vega. In a spring evening it may be seen in the northeast, in the later summer near the zenith, in the autumn in the northwest. On the scale we have laid down with the earth's orbit as a finger-ring, its distance would be some eight or ten miles. The small stars around it in the same constellation are probably ten, twenty, or fifty times as far.

Now, the greatest fact which modern science has brought to light is that our whole solar system, including the sun, with all its planets, is on a journey towards the constellation Lyra. During our whole lives, in all probability during the whole of human history, we have been flying unceasingly towards this beautiful constellation with a speed to which no motion on earth can compare. The speed has recently been determined with a fair degree of certainty, though not with entire exactness; it is about ten miles a second, and therefore not far from three hundred millions of miles a year. But whatever it may be, it is unceasing and unchanging; for us mortals eternal. We are nearer the constellation by five or six hundred miles every minute we live; we are nearer to it now than we were ten years ago by thousands of millions of miles, and every future generation of our race will be nearer than its predecessor by thousands of millions of miles.

When, where, and how, if ever, did this journey begin—when, where, and how, if ever, will it end? This is the greatest of the unsolved problems of astronomy. An astronomer who should watch the heavens for ten thousand years might gather some faint suggestion of an answer, or he might not. All we can do is to seek for some hints by study and comparison with other stars.

The stars are suns. To put it in another way, the sun is one of the stars, and rather a small one at that. If the sun is moving in the way I have described, may not the stars also be in motion, each on a journey of its own through the wilderness of space? To this question astronomy gives an affirmative answer. Most of the stars nearest to us are found to be in motion, some faster than the sun, some more slowly, and the same is doubtless true of all; only the century of accurate observations at our disposal does not show the motion of the distant ones. A given motion seems slower the more distant the moving body; we have to watch a steamship on the horizon some little time to see that she moves at all. Thus it is that the unsolved problem of the motion of our sun is only one branch of a yet more stupendous one: What mean the motions of the stars—how did they begin, and how, if ever, will they end? So far as we can yet see, each star is going straight ahead on its own journey, without regard to its neighbors, if other stars can be so called. Is each describing some vast orbit which, though looking like a straight line during the short period of our observation, will really be seen to curve after ten thousand or a hundred thousand years, or will it go straight on forever? If the laws of motion are true for all space and all time, as we are forced to believe, then each moving star will go on in an unbending line forever unless hindered by the attraction of other stars. If they go on thus, they must, after countless years, scatter in all directions, so that the inhabitants of each shall see only a black, starless sky.

Mathematical science can throw only a few glimmers of light on the questions thus suggested. From what little we know of the masses, distances, and numbers of the stars we see a possibility that the more slow-moving ones may, in long ages, be stopped in their onward courses or brought into orbits of some sort by the attraction of their millions of fellows. But it is hard to admit even this possibility in the case of the swift-moving ones. Attraction, varying as the inverse square of the distance, diminishes so rapidly as the distance increases that, at the distances which separate the stars, it is small indeed. We could not, with the most delicate balance that science has yet invented, even show the attraction of the greatest known star. So far as we know, the two swiftest-moving stars are, first, Arcturus, and, second, one known in astronomy as 1830 Groombridge, the latter so called because it was first observed by the astronomer Groombridge, and is numbered 1830 in his catalogue of stars. If our determinations of the distances of these bodies are to be relied on, the velocity of their motion cannot be much less than two hundred miles a second. They would make the circuit of the earth every two or three minutes. A body massive enough to control this motion would throw a large part of the universe into disorder. Thus the problem where these stars came from and where they are going is for us insoluble, and is all the more so from the fact that the swiftly moving stars are moving in different directions and seem to have no connection with each other or with any known star.

It must not be supposed that these enormous velocities seem so to us. Not one of them, even the greatest, would be visible to the naked eye until after years of watching. On our finger-ring scale, 1830 Groombridge would be some ten miles and Arcturus thirty or forty miles away. Either of them would be moving only two or three feet in a year. To the oldest Assyrian priests Lyra looked much as it does to us to-day. Among the bright and well-known stars Arcturus has the most rapid apparent motion, yet Job himself would not to-day see that its position had changed, unless he had noted it with more exactness than any astronomer of his time.

Another unsolved problem among the greatest which present themselves to the astronomer is that of the size of the universe of stars. We know that several thousand of these bodies are visible to the naked eye; moderate telescopes show us millions; our giant telescopes of the present time, when used as cameras to photograph the heavens, show a number past count, perhaps one hundred millions. Are all these stars only those few which happen to be near us in a universe extending out without end, or do they form a collection of stars outside of which is empty infinite space? In other words, has the universe a boundary? Taken in its widest scope this question must always remain unanswered by us mortals because, even if we should discover a boundary within which all the stars and clusters we ever can know are contained, and outside of which is empty space, still we could never prove that this space is empty out to an infinite distance. Far outside of what we call the universe might still exist other universes which we can never see.

It is a great encouragement to the astronomer that, although he cannot yet set any exact boundary to this universe of ours, he is gathering faint indications that it has a boundary, which his successors not many generations hence may locate so that the astronomer shall include creation itself within his mental grasp. It can be shown mathematically that an infinitely extended system of stars would fill the heavens with a blaze of light like that of the noonday sun. As no such effect is produced, it may be concluded that the universe has a boundary. But this does not enable us to locate the boundary, nor to say how many stars may lie outside the farthest stretches of telescopic vision. Yet by patient research we are slowly throwing light on these points and reaching inferences which, not many years ago, would have seemed forever beyond our powers.

Every one now knows that the Milky Way, that girdle of light which spans the evening sky, is formed of clouds of stars too minute to be seen by the unaided vision. It seems to form the base on which the universe is built and to bind all the stars into a system. It comprises by far the larger number of stars that the telescope has shown to exist. Those we see with the naked eye are almost equally scattered over the sky. But the number which the telescope shows us become more and more condensed in the Milky Way as telescope power is increased. The number of new stars brought out with our greatest power is vastly greater in the Milky Way than in the rest of the sky, so that the former contains a great majority of the stars. What is yet more curious, spectroscopic research has shown that a particular kind of stars, those formed of heated gas, are yet more condensed in the central circle of this band; if they were visible to the naked eye, we should see them encircling the heavens as a narrow girdle forming perhaps the base of our whole system of stars. This arrangement of the gaseous or vaporous stars is one of the most singular facts that modern research has brought to light. It seems to show that these particular stars form a system of their own; but how such a thing can be we are still unable to see.

The question of the form and extent of the Milky Way thus becomes the central one of stellar astronomy. Sir William Herschel began by trying to sound its depths; at one time he thought he had succeeded; but before he died he saw that they were unfathomable with his most powerful telescopes. Even today he would be a bold astronomer who would profess to say with certainty whether the smallest stars we can photograph are at the boundary of the system. Before we decide this point we must have some idea of the form and distance of the cloudlike masses of stars which form our great celestial girdle. A most curious fact is that our solar system seems to be in the centre of this galactic universe, because the Milky Way divides the heavens into two equal parts, and seems equally broad at all points. Were we looking at such a girdle as this from one side or the other, this appearance would not be presented. But let us not be too bold. Perhaps we are the victims of some fallacy, as Ptolemy was when he proved, by what looked like sound reasoning, based on undeniable facts, that this earth of ours stood at rest in the centre of the heavens!

A related problem, and one which may be of supreme importance to the future of our race, is, What is the source of the heat radiated by the sun and stars? We know that life on the earth is dependent on the heat which the sun sends it. If we were deprived of this heat we should in a few days be enveloped in a frost which would destroy nearly all vegetation, and in a few months neither man nor animal would be alive, unless crouching over fires soon to expire for want of fuel. We also know that, at a time which is geologically recent, the whole of New England was covered with a sheet of ice, hundreds or even thousands of feet thick, above which no mountain but Washington raised its head. It is quite possible that a small diminution in the supply of heat sent us by the sun would gradually reproduce the great glacier, and once more make the Eastern States like the pole. But the fact is that observations of temperature in various countries for the last two or three hundred years do not show any change in climate which can be attributed to a variation in the amount of heat received from the sun.

The acceptance of this theory of the heat of those heavenly bodies which shine by their own light—sun, stars, and nebulae—still leaves open a problem that looks insoluble with our present knowledge. What becomes of the great flood of heat and light which the sun and stars radiate into empty space with a velocity of one hundred and eighty thousand miles a second? Only a very small fraction of it can be received by the planets or by other stars, because these are mere points compared with their distance from us. Taking the teaching of our science just as it stands, we should say that all this heat continues to move on through infinite space forever. In a few thousand years it reaches the probable confines of our great universe. But we know of no reason why it should stop here. During the hundreds of millions of years since all our stars began to shine, has the first ray of light and heat kept on through space at the rate of one hundred and eighty thousand miles a second, and will it continue to go on for ages to come? If so, think of its distance now, and think of its still going on, to be forever wasted! Rather say that the problem, What becomes of it? is as yet unsolved.

Thus far I have described the greatest of problems; those which we may suppose to concern the inhabitants of millions of worlds revolving round the stars as much as they concern us. Let us now come down from the starry heights to this little colony where we live, the solar system. Here we have the great advantage of being better able to see what is going on, owing to the comparative nearness of the planets. When we learn that these bodies are like our earth in form, size, and motions, the first question we ask is, Could we fly from planet to planet and light on the surface of each, what sort of scenery would meet our eyes? Mountain, forest, and field, a dreary waste, or a seething caldron larger than our earth? If solid land there is, would we find on it the homes of intelligent beings, the lairs of wild beasts, or no living thing at all? Could we breathe the air, would we choke for breath or be poisoned by the fumes of some noxious gas?

To most of these questions science cannot as yet give a positive answer, except in the case of the moon. Our satellite is so near us that we can see it has no atmosphere and no water, and therefore cannot be the abode of life like ours. The contrast of its eternal deadness with the active life around us is great indeed. Here we have weather of so many kinds that we never tire of talking about it. But on the moon there is no weather at all. On our globe so many things are constantly happening that our thousands of daily journals cannot begin to record them. But on the dreary, rocky wastes of the moon nothing ever happens. So far as we can determine, every stone that lies loose on its surface has lain there through untold ages, unchanged and unmoved.

We cannot speak so confidently of the planets. The most powerful telescopes yet made, the most powerful we can ever hope to make, would scarcely shows us mountains, or lakes, rivers, or fields at a distance of fifty millions of miles. Much less would they show us any works of man. Pointed at the two nearest planets, Venus and Mars, they whet our curiosity more than they gratify it. Especially is this the case with Venus. Ever since the telescope was invented observers have tried to find the time of rotation of this planet on its axis. Some have reached one conclusion, some another, while the wisest have only doubted. The great Herschel claimed that the planet was so enveloped in vapor or clouds that no permanent features could be seen on its surface. The best equipped recent observers think they see faint, shadowy patches, which remain the same from day to day, and which show that the planet always presents the same face to the sun, as the moon does to the earth. Others do not accept this conclusion as proved, believing that these patches may be nothing more than variations of light, shade, and color caused by the reflection of the sun's light at various angles from different parts of the planet.

There is also some mystery about the atmosphere of this planet. When Venus passes nearly between us and the sun, her dark hemisphere is turned towards us, her bright one being always towards the sun. But she is not exactly on a line with the sun except on the very rare occasions of a transit across the sun's disk. Hence, on ordinary occasions, when she seems very near on a line with the sun, we see a very small part of the illuminated hemisphere, which now presents the form of a very thin crescent like the new moon. And this crescent is supposed to be a little broader than it would be if only half the planet were illuminated, and to encircle rather more than half the planet. Now, this is just the effect that would be produced by an atmosphere refracting the sun's light around the edge of the illuminated hemisphere.

The difficulty of observations of this kind is such that the conclusion may be open to doubt. What is seen during transits of Venus over the sun's disk leads to more certain, but yet very puzzling, conclusions. The writer will describe what he saw at the Cape of Good Hope during the transit of December 5, 1882. As the dark planet impinged on the bright sun, it of course cut out a round notch from the edge of the sun. At first, when this notch was small, nothing could be seen of the outline of that part of the planet which was outside the sun. But when half the planet was on the sun, the outline of the part still off the sun was marked by a slender arc of light. A curious fact was that this arc did not at first span the whole outline of the planet, but only showed at one or two points. In a few moments another part of the outline appeared, and then another, until, at last, the arc of light extended around the complete outline. All this seems to show that while the planet has an atmosphere, it is not transparent like ours, but is so filled with mist and clouds that the sun is seen through it only as if shining in a fog.

Not many years ago the planet Mars, which is the next one outside of us, was supposed to have a surface like that of our earth. Some parts were of a dark greenish gray hue; these were supposed to be seas and oceans. Other parts had a bright, warm tint; these were supposed to be the continents. During the last twenty years much has been learned as to how this planet looks, and the details of its surface have been mapped by several observers, using the best telescopes under the most favorable conditions of air and climate. And yet it must be confessed that the result of this labor is not altogether satisfactory. It seems certain that the so-called seas are really land and not water. When it comes to comparing Mars with the earth, we cannot be certain of more than a single point of resemblance. This is that during the Martian winter a white cap, as of snow, is formed over the pole, which partially melts away during the summer. The conclusion that there are oceans whose evaporation forms clouds which give rise to this snow seems plausible. But the telescope shows no clouds, and nothing to make it certain that there is an atmosphere to sustain them. There is no certainty that the white deposit is what we call snow; perhaps it is not formed of water at all. The most careful studies of the surface of this planet, under the best conditions, are those made at the Lowell Observatory at Flagstaff, Arizona. Especially wonderful is the system of so-called canals, first seen by Schiaparelli, but mapped in great detail at Flagstaff. But the nature and meaning of these mysterious lines are still to be discovered. The result is that the question of the real nature of the surface of Mars and of what we should see around us could we land upon it and travel over it are still among the unsolved problems of astronomy.

If this is the case with the nearest planets that we can study, how is it with more distant ones? Jupiter is the only one of these of the condition of whose surface we can claim to have definite knowledge. But even this knowledge is meagre. The substance of what we know is that its surface is surrounded by layers of what look like dense clouds, through which nothing can certainly be seen.

I have already spoken of the heat of the sun and its probable origin. But the question of its heat, though the most important, is not the only one that the sun offers us. What is the sun? When we say that it is a very hot globe, more than a million times as large as the earth, and hotter than any furnace that man can make, so that literally "the elements melt with fervent heat" even at its surface, while inside they are all vaporized, we have told the most that we know as to what the sun really is. Of course we know a great deal about the spots, the rotation of the sun on its axis, the materials of which it is composed, and how its surroundings look during a total eclipse. But all this does not answer our question. There are several mysteries which ingenious men have tried to explain, but they cannot prove their explanations to be correct. One is the cause and nature of the spots. Another is that the shining surface of the sun, the "photosphere," as it is technically called, seems so calm and quiet while forces are acting within it of a magnitude quite beyond our conception. Flames in which our earth and everything on it would be engulfed like a boy's marble in a blacksmith's forge are continually shooting up to a height of tens of thousands of miles. One would suppose that internal forces capable of doing this would break the surface up into billows of fire a thousand miles high; but we see nothing of the kind. The surface of the sun seems almost as placid as a lake.

Yet another mystery is the corona of the sun. This is something we should never have known to exist if the sun were not sometimes totally eclipsed by the dark body of the moon. On these rare occasions the sun is seen to be surrounded by a halo of soft, white light, sending out rays in various directions to great distances. This halo is called the corona, and has been most industriously studied and photographed during nearly every total eclipse for thirty years. Thus we have learned much about how it looks and what its shape is. It has a fibrous, woolly structure, a little like the loose end of a much-worn hempen rope. A certain resemblance has been seen between the form of these seeming fibres and that of the lines in which iron filings arrange themselves when sprinkled on paper over a magnet. It has hence been inferred that the sun has magnetic properties, a conclusion which, in a general way, is supported by many other facts. Yet the corona itself remains no less an unexplained phenomenon.

A phenomenon almost as mysterious as the solar corona is the "zodiacal light," which any one can see rising from the western horizon just after the end of twilight on a clear winter or spring evening. The most plausible explanation is that it is due to a cloud of small meteoric bodies revolving round the sun. We should hardly doubt this explanation were it not that this light has a yet more mysterious appendage, commonly called the Gegenschein, or counter-glow. This is a patch of light in the sky in a direction exactly opposite that of the sun. It is so faint that it can be seen only by a practised eye under the most favorable conditions. But it is always there. The latest suggestion is that it is a tail of the earth, of the same kind as the tail of a comet!

We know that the motions of the heavenly bodies are predicted with extraordinary exactness by the theory of gravitation. When one finds that the exact path of the moon's shadow on the earth during a total eclipse of the sun can be mapped out many years in advance, and that the planets follow the predictions of the astronomer so closely that, if you could see the predicted planet as a separate object, it would look, even in a good telescope, as if it exactly fitted over the real planet, one thinks that here at least is a branch of astronomy which is simply perfect. And yet the worlds themselves show slight deviations in their movements which the astronomer cannot always explain, and which may be due to some hidden cause that, when brought to light, shall lead to conclusions of the greatest importance to our race.

One of these deviations is in the rotation of the earth. Sometimes, for several years at a time, it seems to revolve a little faster, and then again a little slower. The changes are very slight; they can be detected only by the most laborious and refined methods; yet they must have a cause, and we should like to know what that cause is.

The moon shows a similar irregularity of motion. For half a century, perhaps through a whole century, she will go around the earth a little ahead of her regular rate, and then for another half-century or more she will fall behind. The changes are very small; they would never have been seen with the unaided eye, yet they exist. What is their cause? Mathematicians have vainly spent years of study in trying to answer this question.

The orbit of Mercury is found by observations to have a slight motion which mathematicians have vainly tried to explain. For some time it was supposed to be caused by the attraction of an unknown planet between Mercury and the sun, and some were so sure of the existence of this planet that they gave it a name, calling it Vulcan. But of late years it has become reasonably certain that no planet large enough to produce the effect observed can be there. So thoroughly has every possible explanation been sifted out and found wanting, that some astronomers are now inquiring whether the law of gravitation itself may not be a little different from what has always been supposed. A very slight deviation, indeed, would account for the facts, but cautious astronomers want other proofs before regarding the deviation of gravitation as an established fact.

Intelligent men have sometimes inquired how, after devoting so much work to the study of the heavens, anything can remain for astronomers to find out. It is a curious fact that, although they were never learning so fast as at the present day, yet there seems to be more to learn now than there ever was before. Great and numerous as are the unsolved problems of our science, knowledge is now advancing into regions which, a few years ago, seemed inaccessible. Where it will stop none can say.

THE NEW PROBLEMS OF THE UNIVERSE

The achievements of the nineteenth century are still a theme of congratulation on the part of all who compare the present state of the world with that of one hundred years ago. And yet, if we should fancy the most sagacious prophet, endowed with a brilliant imagination, to have set forth in the year 1806 the problems that the century might solve and the things which it might do, we should be surprised to see how few of his predictions had come to pass. He might have fancied aerial navigation and a number of other triumphs of the same class, but he would hardly have had either steam navigation or the telegraph in his picture. In 1856 an article appeared in Harper's Magazine depicting some anticipated features of life in A.D. 3000. We have since made great advances, but they bear little resemblance to what the writer imagined. He did not dream of the telephone, but did describe much that has not yet come to pass and probably never will.

The fact is that, much as the nineteenth century has done, its last work was to amuse itself by setting forth more problems for this century to solve than it has ever itself succeeded in mastering. We should not be far wrong in saying that to-day there are more riddles in the universe than there were before men knew that it contained anything more than the objects they could see.

So far as mere material progress is concerned, it may be doubtful whether anything so epoch-making as the steam-engine or the telegraph is held in store for us by the future. But in the field of purely scientific discovery we are finding a crowd of things of which our philosophy did not dream even ten years ago.

The greatest riddles which the nineteenth century has bequeathed to us relate to subjects so widely separated as the structure of the universe and the structure of atoms of matter. We see more and more of these structures, and we see more and more of unity everywhere, and yet new facts difficult of explanation are being added more rapidly than old facts are being explained.

We all know that the nineteenth century was marked by a separation of the sciences into a vast number of specialties, to the subdivisions of which one could see no end. But the great work of the twentieth century will be to combine many of these specialties. The physical philosopher of the present time is directing his thought to the demonstration of the unity of creation. Astronomical and physical researches are now being united in a way which is bringing the infinitely great and the infinitely small into one field of knowledge. Ten years ago the atoms of matter, of which it takes millions of millions to make a drop of water, were the minutest objects with which science could imagine itself to be concerned. Now a body of experimentalists, prominent among whom stand Professors J. J. Thompson, Becquerel, and Roentgen, have demonstrated the existence of objects so minute that they find their way among and between the atoms of matter as rain-drops do among the buildings of a city. More wonderful yet, it seems likely, although it has not been demonstrated, that these little things, called "corpuscles," play an important part in what is going on among the stars. Whether this be true or not, it is certain that there do exist in the universe emanations of some sort, producing visible effects, the investigation of which the nineteenth century has had to bequeath to the twentieth.

For the purpose of the navigator, the direction of the magnetic needle is invariable in any one place, for months and even years; but when exact scientific observations on it are made, it is found subject to numerous slight changes. The most regular of these consists in a daily change of its direction. It moves one way from morning until noon, and then, late in the afternoon and during the night, turns back again to its original pointing. The laws of this change have been carefully studied from observations, which show that it is least at the equator and larger as we go north into middle latitudes; but no explanation of it resting on an indisputable basis has ever been offered.

Besides these regular changes, there are others of a very irregular character. Every now and then the changes in the direction of the magnet are wider and more rapid than those which occur regularly every day. The needle may move back and forth in a way so fitful as to show the action of some unusual exciting cause. Such movements of the needle are commonly seen when there is a brilliant aurora. This connection shows that a magnetic storm and an aurora must be due to the same or some connected causes.

Those of us who are acquainted with astronomical matters know that the number of spots on the sun goes through a regular cycle of change, having a period of eleven years and one or two months. Now, the curious fact is, when the number and violence of magnetic storms are recorded and compared, it is found that they correspond to the spots on the sun, and go through the same period of eleven years. The conclusion seems almost inevitable: magnetic storms are due to some emanation sent out by the sun, which arises from the same cause that produces the spots. This emanation does not go on incessantly, but only in an occasional way, as storms follow each other on the earth. What is it? Every attempt to detect it has been in vain. Professor Hale, at the Yerkes Observatory, has had in operation from time to time, for several years, his ingenious spectroheliograph, which photographs the sun by a single ray of the spectrum. This instrument shows that violent actions are going on in the sun, which ordinary observation would never lead us to suspect. But it has failed to show with certainty any peculiar emanation at the time of a magnetic storm or anything connected with such a storm.

A mystery which seems yet more impenetrable is associated with the so-called new stars which blaze forth from time to time. These offer to our sight the most astounding phenomena ever presented to the physical philosopher. One hundred years ago such objects offered no mystery. There was no reason to suppose that the Creator of the universe had ceased His functions; and, continuing them, it was perfectly natural that He should be making continual additions to the universe of stars. But the idea that these objects are really new creations, made out of nothing, is contrary to all our modern ideas and not in accord with the observed facts. Granting the possibility of a really new star—if such an object were created, it would be destined to take its place among the other stars as a permanent member of the universe. Instead of this, such objects invariably fade away after a few months, and are changed into something very like an ordinary nebula. A question of transcendent interest is that of the cause of these outbursts. It cannot be said that science has, up to the present time, been able to offer any suggestion not open to question. The most definite one is the collision theory, according to which the outburst is due to the clashing together of two stars, one or both of which might previously have been dark, like a planet. The stars which may be actually photographed probably exceed one hundred millions in number, and those which give too little light to affect the photographic plate may be vastly more numerous than those which do. Dark stars revolve around bright ones in an infinite variety of ways, and complex systems of bodies, the members of which powerfully attract each other, are the rule throughout the universe. Moreover, we can set no limit to the possible number of dark or invisible stars that may be flying through the celestial spaces. While, therefore, we cannot regard the theory of collision as established, it seems to be the only one yet put forth which can lay any claim to a scientific basis. What gives most color to it is the extreme suddenness with which the new stars, so far as has yet been observed, invariably blaze forth. In almost every case it has been only two or three days from the time that the existence of such an object became known until it had attained nearly its full brightness. In fact, it would seem that in the case of the star in Perseus, as in most other cases, the greater part of the outburst took place within the space of twenty-four hours. This suddenness and rapidity is exactly what would be the result of a collision.

The most inexplicable feature of all is the rapid formation of a nebula around this star. In the first photographs of the latter, the appearance presented is simply that of an ordinary star. But, in the course of three or four months, the delicate photographs taken at the Lick Observatory showed that a nebulous light surrounded the star, and was continually growing larger and larger. At first sight, there would seem to be nothing extraordinary in this fact. Great masses of intensely hot vapor, shining by their own light, would naturally be thrown out from the star. Or, if the star had originally been surrounded by a very rare nebulous fog or vapor, the latter would be seen by the brilliant light emitted by the star. On this was based an explanation offered by Kapteyn, which at first seemed very plausible. It was that the sudden wave of light thrown out by the star when it burst forth caused the illumination of the surrounding vapor, which, though really at rest, would seem to expand with the velocity of light, as the illumination reached more and more distant regions of the nebula. This result may be made the subject of exact calculation. The velocity of light is such as would make a circuit of the earth more than seven times in a second. It would, therefore, go out from the star at the rate of a million of miles in between five and six seconds. In the lapse of one of our days, the light would have filled a sphere around the star having a diameter more than one hundred and fifty times the distance of the sun from the earth, and more than five times the dimensions of the whole solar system. Continuing its course and enlarging its sphere day after day, the sight presented to us would have been that of a gradually expanding nebulous mass—a globe of faint light continually increasing in size with the velocity of light.

The first sentiment the reader will feel on this subject is doubtless one of surprise that the distance of the star should be so great as this explanation would imply. Six months after the explosion, the globe of light, as actually photographed, was of a size which would have been visible to the naked eye only as a very minute object in the sky. Is it possible that this minute object could have been thousands of times the dimensions of our solar system?

To see how the question stands from this point of view, we must have some idea of the possible distance of the new star. To gain this idea, we must find some way of estimating distances in the universe. For a reason which will soon be apparent, we begin with the greatest structure which nature offers to the view of man. We all know that the Milky Way is formed of countless stars, too minute to be individually visible to the naked eye. The more powerful the telescope through which we sweep the heavens, the greater the number of the stars that can be seen in it. With the powerful instruments which are now in use for photographing the sky, the number of stars brought to light must rise into the hundreds of millions, and the greater part of these belong to the Milky Way. The smaller the stars we count, the greater their comparative number in the region of the Milky Way. Of the stars visible through the telescope, more than one-half are found in the Milky Way, which may be regarded as a girdle spanning the entire visible universe.

Of the diameter of this girdle we can say, almost with certainty, that it must be more than a thousand times as great as the distance of the nearest fixed star from us, and is probably two or three times greater. According to the best judgment we can form, our solar system is situate near the central region of the girdle, so that the latter must be distant from us by half its diameter. It follows that if we can imagine a gigantic pair of compasses, of which the points extend from us to Alpha Centauri, the nearest star, we should have to measure out at least five hundred spaces with the compass, and perhaps even one thousand or more, to reach the region of the Milky Way.

With this we have to connect another curious fact. Of eighteen new stars which have been observed to blaze forth during the last four hundred years, all are in the region of the Milky Way. This seems to show that, as a rule, they belong to the Milky Way. Accepting this very plausible conclusion, the new star in Perseus must have been more than five hundred times as far as the nearest fixed star. We know that it takes light four years to reach us from Alpha Centauri. It follows that the new star was at a distance through which light would require more than two thousand years to travel, and quite likely a time two or three times this. It requires only the most elementary ideas of geometry to see that if we suppose a ray of light to shoot from a star at such a distance in a direction perpendicular to the line of sight from us to the star, we can compute how fast the ray would seem to us to travel. Granting the distance to be only two thousand light years, the apparent size of the sphere around the star which the light would fill at the end of one year after the explosion would be that of a coin seen at a distance of two thousand times its radius, or one thousand times its diameter—say, a five-cent piece at the distance of sixty feet. But, as a matter of fact, the nebulous illumination expanded with a velocity from ten to twenty times as great as this.

The idea that the nebulosity around the new star was formed by the illumination caused by the light of the explosion spreading out on all sides therefore fails to satisfy us, not because the expansion of the nebula seemed to be so slow, but because it was many times as swift as the speed of light.

Another reason for believing that it was not a mere wave of light is offered by the fact that it did not take place regularly in every direction from the star, but seemed to shoot off at various angles.

Up to the present time, the speed of light has been to science, as well as to the intelligence of our race, almost a symbol of the greatest of possible speeds. The more carefully we reflect on the case, the more clearly we shall see the difficulty in supposing any agency to travel at the rate of the seeming emanations from the new star in Perseus.

As the emanation is seen spreading day after day, the reader may inquire whether this is not an appearance due to some other cause than the mere motion of light. May not an explosion taking place in the centre of a star produce an effect which shall travel yet faster than light? We can only reply that no such agency is known to science.

But is there really anything intrinsically improbable in an agency travelling with a speed many times that of light? In considering that there is, we may fall into an error very much like that into which our predecessors fell in thinking it entirely out of the range of reasonable probability that the stars should be placed at such distances as we now know them to be.

Accepting it as a fact that agencies do exist which travel from sun to planet and from star to star with a speed which beggars all our previous ideas, the first question that arises is that of their nature and mode of action. This question is, up to the present time, one which we do not see any way of completely answering. The first difficulty is that we have no evidence of these agents except that afforded by their action. We see that the sun goes through a regular course of pulsations, each requiring eleven years for completion; and we see that, simultaneously with these, the earth's magnetism goes through a similar course of pulsations. The connection of the two, therefore, seems absolutely proven. But when we ask by what agency it is possible for the sun to affect the magnetism of the earth, and when we trace the passage of some agent between the two bodies, we find nothing to explain the action. To all appearance, the space between the earth and the sun is a perfect void. That electricity cannot of itself pass through a vacuum seems to be a well-established law of physics. It is true that electromagnetic waves, which are supposed to be of the same nature with those of light, and which are used in wireless telegraphy, do pass through a vacuum and may pass from the sun to the earth. But there is no way of explaining how such waves would either produce or affect the magnetism of the earth.

The mysterious emanations from various substances, under certain conditions, may have an intimate relation with yet another of the mysteries of the universe. It is a fundamental law of the universe that when a body emits light or heat, or anything capable of being transformed into light or heat, it can do so only by the expenditure of force, limited in supply. The sun and stars are continually sending out a flood of heat. They are exhausting the internal supply of something which must be limited in extent. Whence comes the supply? How is the heat of the sun kept up? If it were a hot body cooling off, a very few years would suffice for it to cool off so far that its surface would become solid and very soon cold. In recent years, the theory universally accepted has been that the supply of heat is kept up by the continual contraction of the sun, by mutual gravitation of its parts as it cools off. This theory has the advantage of enabling us to calculate, with some approximation to exactness, at what rate the sun must be contracting in order to keep up the supply of heat which it radiates. On this theory, it must, ten millions of years ago, have had twice its present diameter, while less than twenty millions of years ago it could not have existed except as an immense nebula filling the whole solar system. We must bear in mind that this theory is the only one which accounts for the supply of heat, even through human history. If it be true, then the sun, earth, and solar system must be less than twenty million years old.

Here the geologists step in and tell us that this conclusion is wholly inadmissible. The study of the strata of the earth and of many other geological phenomena, they assure us, makes it certain that the earth must have existed much in its present condition for hundreds of millions of years. During all that time there can have been no great diminution in the supply of heat radiated by the sun.

The astronomer, in considering this argument, has to admit that he finds a similar difficulty in connection with the stars and nebulas. It is an impossibility to regard these objects as new; they must be as old as the universe itself. They radiate heat and light year after year. In all probability, they must have been doing so for millions of years. Whence comes the supply? The geologist may well claim that until the astronomer explains this mystery in his own domain, he cannot declare the conclusions of geology as to the age of the earth to be wholly inadmissible.

Now, the scientific experiments of the last two years have brought this mystery of the celestial spaces right down into our earthly laboratories. M. and Madame Curie have discovered the singular metal radium, which seems to send out light, heat, and other rays incessantly, without, so far as has yet been determined, drawing the required energy from any outward source. As we have already pointed out, such an emanation must come from some storehouse of energy. Is the storehouse, then, in the medium itself, or does the latter draw it from surrounding objects? If it does, it must abstract heat from these objects. This question has been settled by Professor Dewar, at the Royal Institution, London, by placing the radium in a medium next to the coldest that art has yet produced—liquid air. The latter is surrounded by the only yet colder medium, liquid hydrogen, so that no heat can reach it. Under these circumstances, the radium still gives out heat, boiling away the liquid air until the latter has entirely disappeared. Instead of the radiation diminishing with time, it rather seems to increase.

Called on to explain all this, science can only say that a molecular change must be going on in the radium, to correspond to the heat it gives out. What that change may be is still a complete mystery. It is a mystery which we find alike in those minute specimens of the rarest of substances under our microscopes, in the sun, and in the vast nebulous masses in the midst of which our whole solar system would be but a speck. The unravelling of this mystery must be the great work of science of the twentieth century. What results shall follow for mankind one cannot say, any more than he could have said two hundred years ago what modern science would bring forth. Perhaps, before future developments, all the boasted achievements of the nineteenth century may take the modest place which we now assign to the science of the eighteenth century—that of the infant which is to grow into a man.

THE STRUCTURE OF THE UNIVERSE

The questions of the extent of the universe in space and of its duration in time, especially of its possible infinity in either space or time, are of the highest interest both in philosophy and science. The traditional philosophy had no means of attacking these questions except considerations suggested by pure reason, analogy, and that general fitness of things which was supposed to mark the order of nature. With modern science the questions belong to the realm of fact, and can be decided only by the results of observation and a study of the laws to which these results may lead.

From the philosophic stand-point, a discussion of this subject which is of such weight that in the history of thought it must be assigned a place above all others, is that of Kant in his "Kritik." Here we find two opposing propositions—the thesis that the universe occupies only a finite space and is of finite duration; the antithesis that it is infinite both as regards extent in space and duration in time. Both of these opposing propositions are shown to admit of demonstration with equal force, not directly, but by the methods of reductio ad absurdum. The difficulty, discussed by Kant, was more tersely expressed by Hamilton in pointing out that we could neither conceive of infinite space nor of space as bounded. The methods and conclusions of modern astronomy are, however, in no way at variance with Kant's reasoning, so far as it extends. The fact is that the problem with which the philosopher of Konigsberg vainly grappled is one which our science cannot solve any more than could his logic. We may hope to gain complete information as to everything which lies within the range of the telescope, and to trace to its beginning every process which we can now see going on in space. But before questions of the absolute beginning of things, or of the boundary beyond which nothing exists, our means of inquiry are quite powerless.

Another example of the ancient method is found in the great work of Copernicus. It is remarkable how completely the first expounder of the system of the world was dominated by the philosophy of his time, which he had inherited from his predecessors. This is seen not only in the general course of thought through the opening chapters of his work, but among his introductory propositions. The first of these is that the universe—mundus—as well as the earth, is spherical in form. His arguments for the sphericity of the earth, as derived from observation, are little more than a repetition of those of Ptolemy, and therefore not of special interest. His proposition that the universe is spherical is, however, not based on observation, but on considerations of the perfection of the spherical form, the general tendency of bodies—a drop of water, for example—to assume this form, and the sphericity of the sun and moon. The idea retained its place in his mind, although the fundamental conception of his system did away with the idea of the universe having any well-defined form.

The question as attacked by modern astronomy is this: we see scattered through space in every direction many millions of stars of various orders of brightness and at distances so great as to defy exact measurement, except in the case of a few of the nearest. Has this collection of stars any well-defined boundary, or is what we see merely that part of an infinite mass which chances to lie within the range of our telescopes? If we were transported to the most distant star of which we have knowledge, should we there find ourselves still surrounded by stars on all sides, or would the space beyond be void? Granting that, in any or every direction, there is a limit to the universe, and that the space beyond is therefore void, what is the form of the whole system and the distance of its boundaries? Preliminary in some sort to these questions are the more approachable ones: Of what sort of matter is the universe formed? and into what sort of bodies is this matter collected?

To the ancients the celestial sphere was a reality, instead of a mere effect of perspective, as we regard it. The stars were set on its surface, or at least at no great distance within its crystalline mass. Outside of it imagination placed the empyrean. When and how these conceptions vanished from the mind of man, it would be as hard to say as when and how Santa Claus gets transformed in the mind of the child. They are not treated as realities by any astronomical writer from Ptolemy down; yet, the impressions and forms of thought to which they gave rise are well marked in Copernicus and faintly evident in Kepler. The latter was perhaps the first to suggest that the sun might be one of the stars; yet, from defective knowledge of the relative brightness of the latter, he was led to the conclusion that their distances from each other were less than the distance which separated them from the sun. The latter he supposed to stand in the centre of a vast vacant region within the system of stars.

For us the great collection of millions of stars which are made known to us by the telescope, together with all the invisible bodies which may be contained within the limits of the system, form the universe. Here the term "universe" is perhaps objectionable because there may be other systems than the one with which we are acquainted. The term stellar system is, therefore, a better one by which to designate the collection of stars in question.

It is remarkable that the first known propounder of that theory of the form and arrangement of the system which has been most generally accepted seems to have been a writer otherwise unknown in science—Thomas Wright, of Durham, England. He is said to have published a book on the theory of the universe, about 1750. It does not appear that this work was of a very scientific character, and it was, perhaps, too much in the nature of a speculation to excite notice in scientific circles. One of the curious features of the history is that it was Kant who first cited Wright's theory, pointed out its accordance with the appearance of the Milky Way, and showed its general reasonableness. But, at the time in question, the work of the philosopher of Konigsberg seems to have excited no more notice among his scientific contemporaries than that of Wright.

Kant's fame as a speculative philosopher has so eclipsed his scientific work that the latter has but recently been appraised at its true value. He was the originator of views which, though defective in detail, embodied a remarkable number of the results of recent research on the structure and form of the universe, and the changes taking place in it. The most curious illustration of the way in which he arrived at a correct conclusion by defective reasoning is found in his anticipation of the modern theory of a constant retardation of the velocity with which the earth revolves on its axis. He conceived that this effect must result from the force exerted by the tidal wave, as moving towards the west it strikes the eastern coasts of Asia and America. An opposite conclusion was reached by Laplace, who showed that the effect of this force was neutralized by forces producing the wave and acting in the opposite direction. And yet, nearly a century later, it was shown that while Laplace was quite correct as regards the general principles involved, the friction of the moving water must prevent the complete neutralization of the two opposing forces, and leave a small residual force acting towards the west and retarding the rotation. Kant's conclusion was established, but by an action different from that which he supposed.

The theory of Wright and Kant, which was still further developed by Herschel, was that our stellar system has somewhat the form of a flattened cylinder, or perhaps that which the earth would assume if, in consequence of more rapid rotation, the bulging out at its equator and the flattening at its poles were carried to an extreme limit. This form has been correctly though satirically compared to that of a grindstone. It rests to a certain extent, but not entirely, on the idea that the stars are scattered through space with equal thickness in every direction, and that the appearance of the Milky Way is due to the fact that we, situated in the centre of this flattened system, see more stars in the direction of the circumference of the system than in that of its poles. The argument on which the view in question rests may be made clear in the following way.

Let us chose for our observations that hour of the night at which the Milky Way skirts our horizon. This is nearly the case in the evenings of May and June, though the coincidence with the horizon can never be exact except to observers stationed near the tropics. Using the figure of the grindstone, we at its centre will then have its circumference around our horizon, while the axis will be nearly vertical. The points in which the latter intersects the celestial sphere are called the galactic poles. There will be two of these poles, the one at the hour in question near the zenith, the other in our nadir, and therefore invisible to us, though seen by our antipodes. Our horizon corresponds, as it were, to the central circle of the Milky Way, which now surrounds us on all sides in a horizontal direction, while the galactic poles are 90 degrees distant from every part of it, as every point of the horizon is 90 degrees from the zenith.

Let us next count the number of stars visible in a powerful telescope in the region of the heavens around the galactic pole, now our zenith, and find the average number per square degree. This will be the richness of the region in stars. Then we take regions nearer the horizontal Milky Way—say that contained between 10 degrees and 20 degrees from the zenith—and, by a similar count, find its richness in stars. We do the same for other regions, nearer and nearer to the horizon, till we reach the galaxy itself. The result of all the counts will be that the richness of the sky in stars is least around the galactic pole, and increases in every direction towards the Milky Way.

Without such counts of the stars we might imagine our stellar system to be a globular collection of stars around which the object in question passed as a girdle; and we might take a globe with a chain passing around it as representative of the possible figure of the stellar system. But the actual increase in star-thickness which we have pointed out shows us that this view is incorrect. The nature and validity of the conclusions to be drawn can be best appreciated by a statement of some features of this tendency of the stars to crowd towards the galactic circle.

Most remarkable is the fact that the tendency is seen even among the brighter stars. Without either telescope or technical knowledge, the careful observer of the stars will notice that the most brilliant constellations show this tendency. The glorious Orion, Canis Major containing the brightest star in the heavens, Cassiopeia, Perseus, Cygnus, and Lyra with its bright-blue Vega, not to mention such constellations as the Southern Cross, all lie in or near the Milky Way. Schiaparelli has extended the investigation to all the stars visible to the naked eye. He laid down on planispheres the number of such stars in each region of the heavens of 5 degrees square. Each region was then shaded with a tint that was darker as the region was richer in stars. The very existence of the Milky Way was ignored in this work, though his most darkly shaded regions lie along the course of this belt. By drawing a band around the sky so as to follow or cover his darkest regions, we shall rediscover the course of the Milky Way without any reference to the actual object. It is hardly necessary to add that this result would be reached with yet greater precision if we included the telescopic stars to any degree of magnitude—plotting them on a chart and shading the chart in the same way. What we learn from this is that the stellar system is not an irregular chaos; and that notwithstanding all its minor irregularities, it may be considered as built up with special reference to the Milky Way as a foundation.

Another feature of the tendency in question is that it is more and more marked as we include fainter stars in our count. The galactic region is perhaps twice as rich in stars visible to the naked eye as the rest of the heavens. In telescopic stars to the ninth magnitude it is three or four times as rich. In the stars found on the photographs of the sky made at the Harvard and other observatories, and in the stargauges of the Herschels, it is from five to ten times as rich.

Another feature showing the unity of the system is the symmetry of the heavens on the two sides of the galactic belt Let us return to our supposition of such a position of the celestial sphere, with respect to the horizon, that the latter coincides with the central line of this belt, one galactic pole being near our zenith. The celestial hemisphere which, being above our horizon, is visible to us, is the one to which we have hitherto directed our attention in describing the distribution of the stars. But below our horizon is another hemisphere, that of our antipodes, which is the counterpart of ours. The stars which it contains are in a different part of the universe from those which we see, and, without unity of plan, would not be subject to the same law. But the most accurate counts of stars that have been made fail to show any difference in their general arrangement in the two hemispheres. They are just as thick around the south galactic poles as around the north one. They show the same tendency to crowd towards the Milky Way in the hemisphere invisible to us as in the hemisphere which we see. Slight differences and irregularities, are, indeed, found in the enumeration, but they are no greater than must necessarily arise from the difficulty of stopping our count at a perfectly fixed magnitude. The aim of star-counts is not to estimate the total number of stars, for this is beyond our power, but the number visible with a given telescope. In such work different observers have explored different parts of the sky, and in a count of the same region by two observers we shall find that, although they attempt to stop at the same magnitude, each will include a great number of stars which the other omits. There is, therefore, room for considerable difference in the numbers of stars recorded, without there being any actual inequality between the two hemispheres.

A corresponding similarity is found in the physical constitution of the stars as brought out by the spectroscope. The Milky Way is extremely rich in bluish stars, which make up a considerable majority of the cloudlike masses there seen. But when we recede from the galaxy on one side, we find the blue stars becoming thinner, while those having a yellow tinge become relatively more numerous. This difference of color also is the same on the two sides of the galactic plane. Nor can any systematic difference be detected between the proper motions of the stars in these two hemispheres. If the largest known proper motion is found in the one, the second largest is in the other. Counting all the known stars that have proper motions exceeding a given limit, we find about as many in one hemisphere as in the other. In this respect, also, the universe appears to be alike through its whole extent. It is the uniformity thus prevailing through the visible universe, as far as we can see, in two opposite directions, which inspires us with confidence in the possibility of ultimately reaching some well-founded conclusion as to the extent and structure of the system.

All these facts concur in supporting the view of Wright, Kant, and Herschel as to the form of the universe. The farther out the stars extend in any direction, the more stars we may see in that direction. In the direction of the axis of the cylinder, the distances of the boundary are least, so that we see fewer stars. The farther we direct our attention towards the equatorial regions of the system, the greater the distance from us to the boundary, and hence the more stars we see. The fact that the increase in the number of stars seen towards the equatorial region of the system is greater, the smaller the stars, is the natural consequence of the fact that distant stars come within our view in greater numbers towards the equatorial than towards the polar regions.

Objections have been raised to the Herschelian view on the ground that it assumes an approximately uniform distribution of the stars in space. It has been claimed that the fact of our seeing more stars in one direction than in another may not arise merely from our looking through a deeper stratum, as Herschel supposed, but may as well be due to the stars being more thinly scattered in the direction of the axis of the system than in that of its equatorial region. The great inequalities in the richness of neighboring regions in the Milky Way show that the hypothesis of uniform distribution does not apply to the equatorial region. The claim has therefore been made that there is no proof of the system extending out any farther in the equatorial than in the polar direction.

The consideration of this objection requires a closer inquiry as to what we are to understand by the form of our system. We have already pointed out the impossibility of assigning any boundary beyond which we can say that nothing exists. And even as regards a boundary of our stellar system, it is impossible for us to assign any exact limit beyond which no star is visible to us. The analogy of collections of stars seen in various parts of the heavens leads us to suppose that there may be no well-defined form to our system, but that, as we go out farther and farther, we shall see occasional scattered stars to, possibly, an indefinite distance. The truth probably is that, as in ascending a mountain, we find the trees, which may be very dense at its base, thin out gradually as we approach the summit, where there may be few or none, so we might find the stars to thin out could we fly to the distant regions of space. The practical question is whether, in such a flight, we should find this sooner by going in the direction of the axis of our system than by directing our course towards the Milky Way. If a point is at length reached beyond which there are but few scattered stars, such a point would, for us, mark the boundary of our system. From this point of view the answer does not seem to admit of doubt. If, going in every direction, we mark the point, if any, at which the great mass of the stars are seen behind us, the totality of all these points will lie on a surface of the general form that Herschel supposed.

There is still another direct indication of the finitude of our stellar system upon which we have not touched. If this system extended out without limit in any direction whatever, it is shown by a geometric process which it is not necessary to explain in the present connection, but which is of the character of mathematical demonstration, that the heavens would, in every direction where this was true, blaze with the light of the noonday sun. This would be very different from the blue-black sky which we actually see on a clear night, and which, with a reservation that we shall consider hereafter, shows that, how far so-ever our stellar system may extend, it is not infinite. Beyond this negative conclusion the fact does not teach us much. Vast, indeed, is the distance to which the system might extend without the sky appearing much brighter than it is, and we must have recourse to other considerations in seeking for indications of a boundary, or even of a well-marked thinning out, of stars.

If, as was formerly supposed, the stars did not greatly differ in the amount of light emitted by each, and if their diversity of apparent magnitude were due principally to the greater distance of the fainter stars, then the brightness of a star would enable us to form a more or less approximate idea of its distance. But the accumulated researches of the past seventy years show that the stars differ so enormously in their actual luminosity that the apparent brightness of a star affords us only a very imperfect indication of its distance. While, in the general average, the brighter stars must be nearer to us than the fainter ones, it by no means follows that a very bright star, even of the first magnitude, is among the nearer to our system. Two stars are worthy of especial mention in this connection, Canopus and Rigel. The first is, with the single exception of Sirius, the brightest star in the heavens. The other is a star of the first magnitude in the southwest corner of Orion. The most long-continued and complete measures of parallax yet made are those carried on by Gill, at the Cape of Good Hope, on these two and some other bright stars. The results, published in 1901, show that neither of these bodies has any parallax that can be measured by the most refined instrumental means known to astronomy. In other words, the distance of these stars is immeasurably great. The actual amount of light emitted by each is certainly thousands and probably tens of thousands of times that of the sun.

Notwithstanding the difficulties that surround the subject, we can at least say something of the distance of a considerable number of the stars. Two methods are available for our estimate—measures of parallax and determination of proper motions.

The problem of stellar parallax, simple though it is in its conception, is the most delicate and difficult of all which the practical astronomer has to encounter. An idea of it may be gained by supposing a minute object on a mountain-top, we know not how many miles away, to be visible through a telescope. The observer is allowed to change the position of his instrument by two inches, but no more. He is required to determine the change in the direction of the object produced by this minute displacement with accuracy enough to determine the distance of the mountain. This is quite analogous to the determination of the change in the direction in which we see a star as the earth, moving through its vast circuit, passes from one extremity of its orbit to the other. Representing this motion on such a scale that the distance of our planet from the sun shall be one inch, we find that the nearest star, on the same scale, will be more than four miles away, and scarcely one out of a million will be at a less distance than ten miles. It is only by the most wonderful perfection both in the heliometer, the instrument principally used for these measures, and in methods of observation, that any displacement at all can be seen even among the nearest stars. The parallaxes of perhaps a hundred stars have been determined, with greater or less precision, and a few hundred more may be near enough for measurement. All the others are immeasurably distant; and it is only by statistical methods based on their proper motions and their probable near approach to equality in distribution that any idea can be gained of their distances.

To form a conception of the stellar system, we must have a unit of measure not only exceeding any terrestrial standard, but even any distance in the solar system. For purely astronomical purposes the most convenient unit is the distance corresponding to a parallax of 1", which is a little more than

200,000 times the sun's distance. But for the purposes of all but the professional astronomer the most convenient unit will be the light-year—that is, the distance through which light would travel in one year. This is equal to the product of 186,000 miles, the distance travelled in one second, by 31,558,000, the number of seconds in a year. The reader who chooses to do so may perform the multiplication for himself. The product will amount to about 63,000 times the distance of the sun.

The nearest star whose distance we know, Alpha Centauri, is distant from us more than four light-years. In all likelihood this is really the nearest star, and it is not at all probable that any other star lies within six light-years. Moreover, if we were transported to this star the probability seems to be that the sun would now be the nearest star to us. Flying to any other of the stars whose parallax has been measured, we should probably find that the average of the six or eight nearest stars around us ranges somewhere between five and seven light-years. We may, in a certain sense, call eight light-years a star-distance, meaning by this term the average of the nearest distances from one star to the surrounding ones.

To put the result of measures of parallax into another form, let us suppose, described around our sun as a centre, a system of concentric spheres each of whose surfaces is at the distance of six light-years outside the sphere next within it. The inner is at the distance of six light-years around the sun. The surface of the second sphere will be twelve light-years away, that of the third eighteen, etc. The volumes of space within each of these spheres will be as the cubes of the diameters. The most likely conclusion we can draw from measures of parallax is that the first sphere will contain, beside the sun at its centre, only Alpha Centauri. The second, twelve light-years away, will probably contain, besides these two, six other stars, making eight in all. The third may contain twenty-one more, making twenty-seven stars within the third sphere, which is the cube of three. Within the fourth would probably be found sixty-four stars, this being the cube of four, and so on.

Beyond this no measures of parallax yet made will give us much assistance. We can only infer that probably the same law holds for a large number of spheres, though it is quite certain that it does not hold indefinitely. For more light on the subject we must have recourse to the proper motions. The latest words of astronomy on this subject may be briefly summarized. As a rule, no star is at rest. Each is moving through space with a speed which differs greatly with different stars, but is nearly always swift, indeed, when measured by any standard to which we are accustomed. Slow and halting, indeed, is that star which does not make more than a mile a second. With two or three exceptions, where the attraction of a companion comes in, the motion of every star, so far as yet determined, takes place in a straight line. In its outward motion the flying body deviates neither to the right nor left. It is safe to say that, if any deviation is to take place, thousands of years will be required for our terrestrial observers to recognize it.

Rapid as the course of these objects is, the distances which we have described are such that, in the great majority of cases, all the observations yet made on the positions of the stars fail to show any well-established motion. It is only in the case of the nearer of these objects that we can expect any motion to be perceptible during the period, in no case exceeding one hundred and fifty years, through which accurate observations extend. The efforts of all the observatories which engage in such work are, up to the present time, unequal to the task of grappling with the motions of all the stars that can be seen with the instruments, and reaching a decision as to the proper motion in each particular case. As the question now stands, the aim of the astronomer is to determine what stars have proper motions large enough to be well established. To make our statement on this subject clear, it must be understood that by this term the astronomer does not mean the speed of a star in space, but its angular motion as he observes it on the celestial sphere. A star moving forward with a given speed will have a greater proper motion according as it is nearer to us. To avoid all ambiguity, we shall use the term "speed" to express the velocity in miles per second with which such a body moves through space, and the term "proper motion" to express the apparent angular motion which the astronomer measures upon the celestial sphere.

Up to the present time, two stars have been found whose proper motions are so large that, if continued, the bodies would make a complete circuit of the heavens in less than 200,000 years. One of these would require about 160,000; the other about 180,000 years for the circuit. Of other stars having a rapid motion only about one hundred would complete their course in less than a million of years.

Quite recently a system of observations upon stars to the ninth magnitude has been nearly carried through by an international combination of observatories. The most important conclusion from these observations relates to the distribution of the stars with reference to the Milky Way, which we have already described. We have shown that stars of every magnitude, bright and faint, show a tendency to crowd towards this belt. It is, therefore, remarkable that no such tendency is seen in the case of those stars which have proper motions large enough to be accurately determined. So far as yet appears, such stars are equally scattered over the heavens, without reference to the course of the Milky Way. The conclusion is obvious. These stars are all inside the girdle of the Milky Way, and within the sphere which contains them the distribution in space is approximately uniform. At least there is no well-marked condensation in the direction of the galaxy nor any marked thinning out towards its poles. What can we say as to the extent of this sphere?

To answer this question, we have to consider whether there is any average or ordinary speed that a star has in space. A great number of motions in the line of sight—that is to say, in the direction of the line from us to the star—have been measured with great precision by Campbell at the Lick Observatory, and by other astronomers. The statistical investigations of Kaptoyn also throw much light on the subject. The results of these investigators agree well in showing an average speed in space—a straight-ahead motion we may call it—of twenty-one miles per second. Some stars may move more slowly than this to any extent; others more rapidly. In two or three cases the speed exceeds one hundred miles per second, but these are quite exceptional. By taking several thousand stars having a given proper motion, we may form a general idea of their average distance, though a great number of them will exceed this average to a considerable extent. The conclusion drawn in this way would be that the stars having an apparent proper motion of 10" per century or more are mostly contained within, or lie not far outside of a sphere whose surface is at a distance from us of 200 light-years. Granting the volume of space which we have shown that nature seems to allow to each star, this sphere should contain 27,000 stars in all. There are about 10,000 stars known to have so large a proper motion as 10". But there is no actual discordance between these results, because not only are there, in all probability, great numbers of stars of which the proper motion is not yet recognized, but there are within the sphere a great number of stars whose motion is less than the average. On the other hand, it is probable that a considerable number of the 10,000 stars lie at a distance at least one-half greater than that of the radius of the sphere.

On the whole, it seems likely that, out to a distance of 300 or even 400 light-years, there is no marked inequality in star distribution. If we should explore the heavens to this distance, we should neither find the beginning of the Milky Way in one direction nor a very marked thinning out in the other. This conclusion is quite accordant with the probabilities of the case. If all the stars which form the groundwork of the Milky Way should be blotted out, we should probably find 100,000,000, perhaps even more, remaining. Assigning to each star the space already shown to be its quota, we should require a sphere of about 3000 light-years radius to contain such a number of stars. At some such distance as this, we might find a thinning out of the stars in the direction of the galactic poles, or the commencement of the Milky Way in the direction of this stream.

Even if this were not found at the distance which we have supposed, it is quite certain that, at some greater distance, we should at least find that the region of the Milky Way is richer in stars than the region near the galactic poles. There is strong reason, based on the appearance of the stars of the Milky Way, their physical constitution, and their magnitudes as seen in the telescope, to believe that, were we placed on one of these stars, we should find the stars around us to be more thickly strewn than they are around our system. In other words, the quota of space filled by each star is probably less in the region of the Milky Way than it is near the centre where we seem to be situated.

We are, therefore, presented with what seems to be the most extraordinary spectacle that the universe can offer, a ring of stars spanning it, and including within its limits by far the great majority of the stars within our system. We have in this spectacle another example of the unity which seems to pervade the system. We might imagine the latter so arranged as to show diversity to any extent. We might have agglomerations of stars like those of the Milky Way situated in some corner of the system, or at its centre, or scattered through it here and there in every direction. But such is not the case. There are, indeed, a few star-clusters scattered here and there through the system; but they are essentially different from the clusters of the Milky Way, and cannot be regarded as forming an important part of the general plan. In the case of the galaxy we have no such scattering, but find the stars built, as it were, into this enormous ring, having similar characteristics throughout nearly its whole extent, and having within it a nearly uniform scattering of stars, with here and there some collected into clusters. Such, to our limited vision, now appears the universe as a whole.

We have already alluded to the conclusion that an absolutely infinite system of stars would cause the entire heavens to be filled with a blaze of light as bright as the sun. It is also true that the attractive force within such a universe would be infinitely great in some direction or another. But neither of these considerations enables us to set a limit to the extent of our system. In two remarkable papers by Lord Kelvin which have recently appeared, the one being an address before the British Association at its Glasgow meeting, in 1901, are given the results of some numerical computations pertaining to this subject. Granting that the stars are scattered promiscuously through space with some approach to uniformity in thickness, and are of a known degree of brilliancy, it is easy to compute how far out the system must extend in order that, looking up at the sky, we shall see a certain amount of light coming from the invisible stars. Granting that, in the general average, each star is as bright as the sun, and that their thickness is such that within a sphere of 3300 light-years there are 1,000,000,000 stars, if we inquire how far out such a system must be continued in order that the sky shall shine with even four per cent of the light of the sun, we shall find the distance of its boundary so great that millions of millions of years would be required for the light of the outer stars to reach the centre of the system. In view of the fact that this duration in time far exceeds what seems to be the possible life duration of a star, so far as our knowledge of it can extend, the mere fact that the sky does not glow with any such brightness proves little or nothing as to the extent of the system.

We may, however, replace these purely negative considerations by inquiring how much light we actually get from the invisible stars of our system. Here we can make a definite statement. Mark out a small circle in the sky 1 degree in diameter. The quantity of light which we receive on a cloudless and moonless night from the sky within this circle admits of actual determination. From the measures so far available it would seem that, in the general average, this quantity of light is not very different from that of a star of the fifth magnitude. This is something very different from a blaze of light. A star of the fifth magnitude is scarcely more than plainly visible to ordinary vision. The area of the whole sky is, in round numbers, about 50,000 times that of the circle we have described. It follows that the total quantity of light which we receive from all the stars is about equal to that of 50,000 stars of the fifth magnitude—somewhat more than 1000 of the first magnitude. This whole amount of light would have to be multiplied by 90,000,000 to make a light equal to that of the sun. It is, therefore, not at all necessary to consider how far the system must extend in order that the heavens should blaze like the sun. Adopting Lord Kelvin's hypothesis, we shall find that, in order that we may receive from the stars the amount of light we have designated, this system need not extend beyond some 5000 light-years. But this hypothesis probably overestimates the thickness of the stars in space. It does not seem probable that there are as many as 1,000,000,000 stars within the sphere of 3300 light-years. Nor is it at all certain that the light of the average star is equal to that of the sun. It is impossible, in the present state of our knowledge, to assign any definite value to this average. To do so is a problem similar to that of assigning an average weight to each component of the animal creation, from the microscopic insects which destroy our plants up to the elephant. What we can say with a fair approximation to confidence is that, if we could fly out in any direction to a distance of 20,000, perhaps even of 10,000, light-years, we should find that we had left a large fraction of our system behind us. We should see its boundary in the direction in which we had travelled much more certainly than we see it from our stand-point.

We should not dismiss this branch of the subject without saying that considerations are frequently adduced by eminent authorities which tend to impair our confidence in almost any conclusion as to the limits of the stellar system. The main argument is based on the possibility that light is extinguished in its passage through space; that beyond a certain distance we cannot see a star, however bright, because its light is entirely lost before reaching us. That there could be any loss of light in passing through an absolute vacuum of any extent cannot be admitted by the physicist of to-day without impairing what he considers the fundamental principles of the vibration of light. But the possibility that the celestial spaces are pervaded by matter which might obstruct the passage of light is to be considered. We know that minute meteoric particles are flying through our system in such numbers that the earth encounters several millions of them every day, which appear to us in the familiar phenomena of shooting-stars. If such particles are scattered through all space, they must ultimately obstruct the passage of light. We know little of the size of these bodies, but, from the amount of energy contained in their light as they are consumed in the passage through our atmosphere, it does not seem at all likely that they are larger than grains of sand or, perhaps, minute pebbles. They are probably vastly more numerous in the vicinity of the sun than in the interstellar spaces, since they would naturally tend to be collected by the sun's attraction. In fact there are some reasons for believing that most of these bodies are the debris of comets; and the latter are now known to belong to the solar system, and not to the universe at large.

But whatever view we take of these possibilities, they cannot invalidate our conclusion as to the general structure of the stellar system as we know it. Were meteors so numerous as to cut off a large fraction of the light from the more distant stars, we should see no Milky Way, but the apparent thickness of the stars in every direction would be nearly the same. The fact that so many more of these objects are seen around the galactic belt than in the direction of its poles shows that, whatever extinction light may suffer in going through the greatest distances, we see nearly all that comes from stars not more distant than the Milky Way itself.

Intimately connected with the subject we have discussed is the question of the age of our system, if age it can be said to have. In considering this question, the simplest hypothesis to suggest itself is that the universe has existed forever in some such form as we now see it; that it is a self-sustaining system, able to go on forever with only such cycles of transformation as may repeat themselves indefinitely, and may, therefore, have repeated themselves indefinitely in the past. Ordinary observation does not make anything known to us which would seem to invalidate this hypothesis. In looking upon the operations of the universe, we may liken ourselves to a visitor to the earth from another sphere who has to draw conclusions about the life of an individual man from observations extending through a few days. During that time, he would see no reason why the life of the man should have either a beginning or an end. He sees a daily round of change, activity and rest, nutrition and waste; but, at the end of the round, the individual is seemingly restored to his state of the day before. Why may not this round have been going on forever, and continue in the future without end? It would take a profounder course of observation and a longer time to show that, notwithstanding this seeming restoration, an imperceptible residual of vital energy, necessary to the continuance of life, has not been restored, and that the loss of this residuum day by day must finally result in death.

The case is much the same with the great bodies of the universe. Although, to superficial observation, it might seem that they could radiate their light forever, the modern generalizations of physics show that such cannot be the case. The radiation of light necessarily involves a corresponding loss of heat and with it the expenditure of some form of energy. The amount of energy within any body is necessarily limited. The supply must be exhausted unless the energy of the light sent out into infinite space is, in some way, restored to the body which expended it. The possibility of such a restoration completely transcends our science. How can the little vibration which strikes our eye from some distant star, and which has been perhaps thousands of years in reaching us, find its way back to its origin? The light emitted by the sun 10,000 years ago is to-day pursuing its way in a sphere whose surface is 10,000 light-years distant on all sides. Science has nothing even to suggest the possibility of its restoration, and the most delicate observations fail to show any return from the unfathomable abyss.

Up to the time when radium was discovered, the most careful investigations of all conceivable sources of supply had shown only one which could possibly be of long duration. This is the contraction which is produced in the great incandescent bodies of the universe by the loss of the heat which they radiate. As remarked in the preceding essay, the energy generated by the sun's contraction could not have kept up its present supply of heat for much more than twenty or thirty millions of years, while the study of earth and ocean shows evidence of the action of a series of causes which must have been going on for hundreds of millions of years.

The antagonism between the two conclusions is even more marked than would appear from this statement. The period of the sun's heat set by the astronomical physicist is that during which our luminary could possibly have existed in its present form. The period set by the geologist is not merely that of the sun's existence, but that during which the causes effecting geological changes have not undergone any complete revolution. If, at any time, the sun radiated much less than its present amount of heat, no water could have existed on the earth's surface except in the form of ice; there would have been scarcely any evaporation, and the geological changes due to erosion could not have taken place. Moreover, the commencement of the geological operations of which we speak is by no means the commencement of the earth's existence. The theories of both parties agree that, for untold aeons before the geological changes now visible commenced, our planet was a molten mass, perhaps even an incandescent globe like the sun. During all those aeons the sun must have been in existence as a vast nebulous mass, first reaching as far as the earth's orbit, and slowly contracting its dimensions. And these aeons are to be included in any estimate of the age of the sun.

The doctrine of cosmic evolution—the theory which in former times was generally known as the nebular hypothesis—that the heavenly bodies were formed by the slow contraction of heated nebulous masses, is indicated by so many facts that it seems scarcely possible to doubt it except on the theory that the laws of nature were, at some former time, different from those which we now see in operation. Granting the evolutionary hypothesis, every star has its lifetime. We can even lay down the law by which it passes from infancy to old age. All stars do not have the same length of life; the rule is that the larger the star, or the greater the mass of matter which composes it, the longer will it endure. Up to the present time, science can do nothing more than point out these indications of a beginning, and their inevitable consequence, that there is to be an end to the light and heat of every heavenly body. But no cautious thinker can treat such a subject with the ease of ordinary demonstration. The investigator may even be excused if he stands dumb with awe before the creation of his own intellect. Our accurate records of the operations of nature extend through only two or three centuries, and do not reach a satisfactory standard until within a single century. The experience of the individual is limited to a few years, and beyond this period he must depend upon the records of his ancestors. All his knowledge of the laws of nature is derived from this very limited experience. How can he essay to describe what may have been going on hundreds of millions of years in the past? Can he dare to say that nature was the same then as now?

It is a fundamental principle of the theory of evolution, as developed by its greatest recent expounder, that matter itself is eternal, and that all the changes which have taken place in the universe, so far as made up of matter, are in the nature of transformations of this eternal substance. But we doubt whether any physical philosopher of the present day would be satisfied to accept any demonstration of the eternity of matter. All he would admit is that, so far as his observation goes, no change in the quantity of matter can be produced by the action of any known cause. It seems to be equally uncreatable and indestructible. But he would, at the same time, admit that his experience no more sufficed to settle the question than the observation of an animal for a single day would settle the question of the duration of its life, or prove that it had neither beginning nor end. He would probably admit that even matter itself may be a product of evolution. The astronomer finds it difficult to conceive that the great nebulous masses which he sees in the celestial spaces—millions of times larger than the whole solar system, yet so tenuous that they offer not the slightest obstruction to the passage of a ray of light through their whole length— situated in what seems to be a region of eternal cold, below anything that we can produce on the earth's surface, yet radiating light, and with it heat, like an incandescent body—can be made up of the same kind of substance that we have around us on the earth's surface. Who knows but that the radiant property that Becquerel has found in certain forms of matter may be a residuum of some original form of energy which is inherent in great cosmical masses, and has fed our sun during all the ages required by the geologist for the structure of the earth's crusts? It may be that in this phenomenon we have the key to the great riddle of the universe, with which profounder secrets of matter than any we have penetrated will be opened to the eyes of our successors.

THE EXTENT OF THE UNIVERSE

We cannot expect that the wisest men of our remotest posterity, who can base their conclusions upon thousands of years of accurate observation, will reach a decision on this subject without some measure of reserve. Such being the case, it might appear the dictate of wisdom to leave its consideration to some future age, when it may be taken up with better means of information than we now possess. But the question is one which will refuse to be postponed so long as the propensity to think of the possibilities of creation is characteristic of our race. The issue is not whether we shall ignore the question altogether, like Eve in the presence of Raphael; but whether in studying it we shall confine our speculations within the limits set by sound scientific reasoning. Essaying to do this, I invite the reader's attention to what science may suggest, admitting in advance that the sphere of exact knowledge is small compared with the possibilities of creation, and that outside this sphere we can state only more or less probable conclusions.

The reader who desires to approach this subject in the most receptive spirit should begin his study by betaking himself on a clear, moonless evening, when he has no earthly concern to disturb the serenity of his thoughts, to some point where he can lie on his back on bench or roof, and scan the whole vault of heaven at one view. He can do this with the greatest pleasure and profit in late summer or autumn—winter would do equally well were it possible for the mind to rise so far above bodily conditions that the question of temperature should not enter. The thinking man who does this under circumstances most favorable for calm thought will form a new conception of the wonder of the universe. If summer or autumn be chosen, the stupendous arch of the Milky Way will pass near the zenith, and the constellation Lyra, led by its beautiful blue Vega of the first magnitude, may be not very far from that point. South of it will be seen the constellation Aquila, marked by the bright Altair, between two smaller but conspicuous stars. The bright Arcturus will be somewhere in the west, and, if the observation is not made too early in the season, Aldebaran will be seen somewhere in the east. When attention is concentrated on the scene the thousands of stars on each side of the Milky Way will fill the mind with the consciousness of a stupendous and all-embracing frame, beside which all human affairs sink into insignificance. A new idea will be formed of such a well-known fact of astronomy as the motion of the solar system in space, by reflecting that, during all human history, the sun, carrying the earth with it, has been flying towards a region in or just south of the constellation Lyra, with a speed beyond all that art can produce on earth, without producing any change apparent to ordinary vision in the aspect of the constellation. Not only Lyra and Aquila, but every one of the thousand stars which form the framework of the sky, were seen by our earliest ancestors just as we see them now. Bodily rest may be obtained at any time by ceasing from our labors, and weary systems may find nerve rest at any summer resort; but I know of no way in which complete rest can be obtained for the weary soul—in which the mind can be so entirely relieved of the burden of all human anxiety—as by the contemplation of the spectacle presented by the starry heavens under the conditions just described. As we make a feeble attempt to learn what science can tell us about the structure of this starry frame, I hope the reader will allow me to at least fancy him contemplating it in this way.

The first question which may suggest itself to the inquiring reader is: How is it possible by any methods of observation yet known to the astronomer to learn anything about the universe as a whole? We may commence by answering this question in a somewhat comprehensive way. It is possible only because the universe, vast though it is, shows certain characteristics of a unified and bounded whole. It is not a chaos, it is not even a collection of things, each of which came into existence in its own separate way. If it were, there would be nothing in common between two widely separate regions of the universe. But, as a matter of fact, science shows unity in the whole structure, and diversity only in details. The Milky Way itself will be seen by the most ordinary observer to form a single structure. This structure is, in some sort, the foundation on which the universe is built. It is a girdle which seems to span the whole of creation, so far as our telescopes have yet enabled us to determine what creation is; and yet it has elements of similarity in all its parts. What has yet more significance, it is in some respects unlike those parts of the universe which lie without it, and even unlike those which lie in that central region within it where our system is now situated. The minute stars, individually far beyond the limit of visibility to the naked eye, which form its cloudlike agglomerations, are found to be mostly bluer in color, from one extreme to the other, than the general average of the stars which make up the rest of the universe.

In the preceding essay on the structure of the universe, we have pointed out several features of the universe showing the unity of the whole. We shall now bring together these and other features with a view of showing their relation to the question of the extent of the universe.

The Milky Way being in a certain sense the foundation on which the whole system is constructed, we have first to notice the symmetry of the whole. This is seen in the fact that a certain resemblance is found in any two opposite regions of the sky, no matter where we choose them. If we take them in the Milky Way, the stars are more numerous than elsewhere; if we take opposite regions in or near the Milky Way, we shall find more stars in both of them than elsewhere; if we take them in the region anywhere around the poles of the Milky Way, we shall find fewer stars, but they will be equally numerous in each of the two regions. We infer from this that whatever cause determined the number of the stars in space was of the same nature in every two antipodal regions of the heavens.

Another unity marked with yet more precision is seen in the chemical elements of which stars are composed. We know that the sun is composed of the same elements which we find on the earth and into which we resolve compounds in our laboratories. These same elements are found in the most distant stars. It is true that some of these bodies seem to contain elements which we do not find on earth. But as these unknown elements are scattered from one extreme of the universe to the other, they only serve still further to enforce the unity which runs through the whole. The nebulae are composed, in part at least, of forms of matter dissimilar to any with which we are acquainted. But, different though they may be, they are alike in their general character throughout the whole field we are considering. Even in such a feature as the proper motions of the stars, the same unity is seen. The reader doubtless knows that each of these objects is flying through space on its own course with a speed comparable with that of the earth around the sun. These speeds range from the smallest limit up to more than one hundred miles a second. Such diversity might seem to detract from the unity of the whole; but when we seek to learn something definite by taking their average, we find this average to be, so far as can yet be determined, much the same in opposite regions of the universe. Quite recently it has become probable that a certain class of very bright stars known as Orion stars—because there are many of them in the most brilliant of our constellations—which are scattered along the whole course of the Milky Way, have one and all, in the general average, slower motions than other stars. Here again we have a definable characteristic extending through the universe. In drawing attention to these points of similarity throughout the whole universe, it must not be supposed that we base our conclusions directly upon them. The point they bring out is that the universe is in the nature of an organized system; and it is upon the fact of its being such a system that we are able, by other facts, to reach conclusions as to its structure, extent, and other characteristics.

One of the great problems connected with the universe is that of its possible extent. How far away are the stars? One of the unities which we have described leads at once to the conclusion that the stars must be at very different distances from us; probably the more distant ones are a thousand times as far as the nearest; possibly even farther than this. This conclusion may, in the first place, be based on the fact that the stars seem to be scattered equally throughout those regions of the universe which are not connected with the Milky Way. To illustrate the principle, suppose a farmer to sow a wheat-field of entirely unknown extent with ten bushels of wheat. We visit the field and wish to have some idea of its acreage. We may do this if we know how many grains of wheat there are in the ten bushels. Then we examine a space two or three feet square in any part of the field and count the number of grains in that space. If the wheat is equally scattered over the whole field, we find its extent by the simple rule that the size of the field bears the same proportion to the size of the space in which the count was made that the whole number of grains in the ten bushels sown bears to the number of grains counted. If we find ten grains in a square foot, we know that the number of square feet in the whole field is one-tenth that of the number of grains sown. So it is with the universe of stars. If the latter are sown equally through space, the extent of the space occupied must be proportional to the number of stars which it contains.

But this consideration does not tell us anything about the actual distance of the stars or how thickly they may be scattered. To do this we must be able to determine the distance of a certain number of stars, just as we suppose the farmer to count the grains in a certain small extent of his wheat-field. There is only one way in which we can make a definite measure of the distance of any one star. As the earth swings through its vast annual circuit round the sun, the direction of the stars must appear to be a little different when seen from one extremity of the circuit than when seen from the other. This difference is called the parallax of the stars; and the problem of measuring it is one of the most delicate and difficult in the whole field of practical astronomy.

The nineteenth century was well on its way before the instruments of the astronomer were brought to such perfection as to admit of the measurement. From the time of Copernicus to that of Bessel many attempts had been made to measure the parallax of the stars, and more than once had some eager astronomer thought himself successful. But subsequent investigation always showed that he had been mistaken, and that what he thought was the effect of parallax was due to some other cause, perhaps the imperfections of his instrument, perhaps the effect of heat and cold upon it or upon the atmosphere through which he was obliged to observe the star, or upon the going of his clock. Thus things went on until 1837, when Bessel announced that measures with a heliometer—the most refined instrument that has ever been used in measurement—showed that a certain star in the constellation Cygnus had a parallax of one-third of a second. It may be interesting to give an idea of this quantity. Suppose one's self in a house on top of a mountain looking out of a window one foot square, at a house on another mountain one hundred miles away. One is allowed to look at that distant house through one edge of the pane of glass and then through the opposite edge; and he has to determine the change in the direction of the distant house produced by this change of one foot in his own position. From this he is to estimate how far off the other mountain is. To do this, one would have to measure just about the amount of parallax that Bessel found in his star. And yet this star is among the few nearest to our system. The nearest star of all, Alpha Centauri, visible only in latitudes south of our middle ones, is perhaps half as far as Bessel's star, while Sirius and one or two others are nearly at the same distance. About 100 stars, all told, have had their parallax measured with a greater or less degree of probability. The work is going on from year to year, each successive astronomer who takes it up being able, as a general rule, to avail himself of better instruments or to use a better method. But, after all, the distances of even some of the 100 stars carefully measured must still remain quite doubtful.

Let us now return to the idea of dividing the space in which the universe is situated into concentric spheres drawn at various distances around our system as a centre. Here we shall take as our standard a distance 400,000 times that of the sun from the earth. Regarding this as a unit, we imagine ourselves to measure out in any direction a distance twice as great as this— then another equal distance, making one three times as great, and so indefinitely. We then have successive spheres of which we take the nearer one as the unit. The total space filled by the second sphere will be 8 times the unit; that of the third space 27 times, and so on, as the cube of each distance. Since each sphere includes all those within it, the volume of space between each two spheres will be proportional to the difference of these numbers—that is, to 1, 7, 19, etc. Comparing these volumes with the number of stars probably within them, the general result up to the present time is that the number of stars in any of these spheres will be about equal to the units of volume which they comprise, when we take for this unit the smallest and innermost of the spheres, having a radius 400,000 times the sun's distance. We are thus enabled to form some general idea of how thickly the stars are sown through space. We cannot claim any numerical exactness for this idea, but in the absence of better methods it does afford us some basis for reasoning.

Now we can carry on our computation as we supposed the farmer to measure the extent of his wheat-field. Let us suppose that there are 125,000,000 stars in the heavens. This is an exceedingly rough estimate, but let us make the supposition for the time being. Accepting the view that they are nearly equally scattered throughout space, it will follow that they must be contained within a volume equal to 125,000,000 times the sphere we have taken as our unit. We find the distance of the surface of this sphere by extracting the cube root of this number, which gives us 500. We may, therefore, say, as the result of a very rough estimate, that the number of stars we have supposed would be contained within a distance found by multiplying 400,000 times the distance of the sun by 500; that is, that they are contained within a region whose boundary is 200,000,000 times the distance of the sun. This is a distance through which light would travel in about 3300 years.

It is not impossible that the number of stars is much greater than that we have supposed. Let us grant that there are eight times as many, or 1,000,000,000. Then we should have to extend the boundary of our universe twice as far, carrying it to a distance which light would require 6600 years to travel.

There is another method of estimating the thickness with which stars are sown through space, and hence the extent of the universe, the result of which will be of interest. It is based on the proper motion of the stars. One of the greatest triumphs of astronomy of our time has been the measurement of the actual speed at which many of the stars are moving to or from us in space. These measures are made with the spectroscope. Unfortunately, they can be best made only on the brighter stars—becoming very difficult in the case of stars not plainly visible to the naked eye. Still the motions of several hundreds have been measured and the number is constantly increasing.

A general result of all these measures and of other estimates may be summed up by saying that there is a certain average speed with which the individual stars move in space; and that this average is about twenty miles per second. We are also able to form an estimate as to what proportion of the stars move with each rate of speed from the lowest up to a limit which is probably as high as 150 miles per second. Knowing these proportions we have, by observation of the proper motions of the stars, another method of estimating how thickly they are scattered in space; in other words, what is the volume of space which, on the average, contains a single star. This method gives a thickness of the stars greater by about twenty-five per cent, than that derived from the measures of parallax. That is to say, a sphere like the second we have proposed, having a radius 800,000 times the distance of the sun, and therefore a diameter 1,600,000 times this distance, would, judging by the proper motions, have ten or twelve stars contained within it, while the measures of parallax only show eight stars within the sphere of this diameter having the sun as its centre. The probabilities are in favor of the result giving the greater thickness of the stars. But, after all, the discrepancy does not change the general conclusion as to the limits of the visible universe. If we cannot estimate its extent with the same certainty that we can determine the size of the earth, we can still form a general idea of it.

The estimates we have made are based on the supposition that the stars are equally scattered in space. We have good reason to believe that this is true of all the stars except those of the Milky Way. But, after all, the latter probably includes half the whole number of stars visible with a telescope, and the question may arise whether our results are seriously wrong from this cause. This question can best be solved by yet another method of estimating the average distance of certain classes of stars.

The parallaxes of which we have heretofore spoken consist in the change in the direction of a star produced by the swing of the earth from one side of its orbit to the other. But we have already remarked that our solar system, with the earth as one of its bodies, has been journeying straightforward through space during all historic times. It follows, therefore, that we are continually changing the position from which we view the stars, and that, if the latter were at rest, we could, by measuring the apparent speed with which they are moving in the opposite direction from that of the earth, determine their distance. But since every star has its own motion, it is impossible, in any one case, to determine how much of the apparent motion is due to the star itself, and how much to the motion of the solar system through space. Yet, by taking general averages among groups of stars, most of which are probably near each other, it is possible to estimate the average distance by this method. When an attempt is made to apply it, so as to obtain a definite result, the astronomer finds that the data now available for the purpose are very deficient. The proper motion of a star can be determined only by comparing its observed position in the heavens at two widely separate epochs. Observations of sufficient precision for this purpose were commenced about 1750 at the Greenwich Observatory, by Bradley, then Astronomer Royal of England. But out of 3000 stars which he determined, only a few are available for the purpose. Even since his time, the determinations made by each generation of astronomers have not been sufficiently complete and systematic to furnish the material for anything like a precise determination of the proper motions of stars. To determine a single position of any one star involves a good deal of computation, and if we reflect that, in order to attack the problem in question in a satisfactory way, we should have observations of 1,000,000 of these bodies made at intervals of at least a considerable fraction of a century, we see what an enormous task the astronomers dealing with this problem have before them, and how imperfect must be any determination of the distance of the stars based on our motion through space. So far as an estimate can be made, it seems to agree fairly well with the results obtained by the other methods. Speaking roughly, we have reason, from the data so far available, to believe that the stars of the Milky Way are situated at a distance between 100,000,000 and 200,000,000 times the distance of the sun. At distances less than this it seems likely that the stars are distributed through space with some approach to uniformity. We may state as a general conclusion, indicated by several methods of making the estimate, that nearly all the stars which we can see with our telescopes are contained within a sphere not likely to be much more than 200,000,000 times the distance of the sun.

The inquiring reader may here ask another question. Granting that all the stars we can see are contained within this limit, may there not be any number of stars outside the limit which are invisible only because they are too far away to be seen?

This question may be answered quite definitely if we grant that light from the most distant stars meets with no obstruction in reaching us. The most conclusive answer is afforded by the measure of starlight. If the stars extended out indefinitely, then the number of those of each order of magnitude would be nearly four times that of the magnitude next brighter. For example, we should have nearly four times as many stars of the sixth magnitude as of the fifth; nearly four times as many of the seventh as of the sixth, and so on indefinitely. Now, it is actually found that while this ratio of increase is true for the brighter stars, it is not so for the fainter ones, and that the increase in the number of the latter rapidly falls off when we make counts of the fainter telescopic stars. In fact, it has long been known that, were the universe infinite in extent, and the stars equally scattered through all space, the whole heavens would blaze with the light of countless millions of distant stars separately invisible even with the telescope.

The only way in which this conclusion can be invalidated is by the possibility that the light of the stars is in some way extinguished or obstructed in its passage through space. A theory to this effect was propounded by Struve nearly a century ago, but it has since been found that the facts as he set them forth do not justify the conclusion, which was, in fact, rather hypothetical. The theories of modern science converge towards the view that, in the pure ether of space, no single ray of light can ever be lost, no matter how far it may travel. But there is another possible cause for the extinction of light. During the last few years discoveries of dark and therefore invisible stars have been made by means of the spectroscope with a success which would have been quite incredible a very few years ago, and which, even to-day, must excite wonder and admiration. The general conclusion is that, besides the shining stars which exist in space, there may be any number of dark ones, forever invisible in our telescopes. May it not be that these bodies are so numerous as to cut off the light which we would otherwise receive from the more distant bodies of the universe? It is, of course, impossible to answer this question in a positive way, but the probable conclusion is a negative one. We may say with certainty that dark stars are not so numerous as to cut off any important part of the light from the stars of the Milky Way, because, if they did, the latter would not be so clearly seen as it is. Since we have reason to believe that the Milky Way comprises the more distant stars of our system, we may feel fairly confident that not much light can be cut off by dark bodies from the most distant region to which our telescopes can penetrate. Up to this distance we see the stars just as they are. Even within the limit of the universe as we understand it, it is likely that more than one-half the stars which actually exist are too faint to be seen by human vision, even when armed with the most powerful telescopes. But their invisibility is due only to their distance and the faintness of their intrinsic light, and not to any obstructing agency.

The possibility of dark stars, therefore, does not invalidate the general conclusions at which our survey of the subject points. The universe, so far as we can see it, is a bounded whole. It is surrounded by an immense girdle of stars, which, to our vision, appears as the Milky Way. While we cannot set exact limits to its distance, we may yet confidently say that it is bounded. It has uniformities running through its vast extent. Could we fly out to distances equal to that of the Milky Way, we should find comparatively few stars beyond the limits of that girdle. It is true that we cannot set any definite limit and say that beyond this nothing exists. What we can say is that the region containing the visible stars has some approximation to a boundary. We may fairly anticipate that each successive generation of astronomers, through coming centuries, will obtain a little more light on the subject—will be enabled to make more definite the boundaries of our system of stars, and to draw more and more probable conclusions as to the existence or non-existence of any object outside of it. The wise investigator of to-day will leave to them the task of putting the problem into a more positive shape.

MAKING AND USING A TELESCOPE

The impression is quite common that satisfactory views of the heavenly bodies can be obtained only with very large telescopes, and that the owner of a small one must stand at a great disadvantage alongside of the fortunate possessor of a great one. This is not true to the extent commonly supposed. Sir William Herschel would have been delighted to view the moon through what we should now consider a very modest instrument; and there are some objects, especially the moon, which commonly present a more pleasing aspect through a small telescope than through a large one. The numerous owners of small telescopes throughout the country might find their instruments much more interesting than they do if they only knew what objects were best suited to examination with the means at their command. There are many others, not possessors of telescopes, who would like to know how one can be acquired, and to whom hints in this direction will be valuable. We shall therefore give such information as we are able respecting the construction of a telescope, and the more interesting celestial objects to which it may be applied.

Whether the reader does or does not feel competent to undertake the making of a telescope, it may be of interest to him to know how it is done. First, as to the general principles involved, it is generally known that the really vital parts of the telescope, which by their combined action perform the office of magnifying the object looked at, are two in number, the OBJECTIVE and the EYE-PIECE. The former brings the rays of light which emanate from the object to the focus where the image of the object is formed. The eye-piece enables the observer to see this image to the best advantage.

The functions of the objective as well as those of the eye-piece may, to a certain extent, each be performed by a single lens. Galileo and his contemporaries made their telescopes in this way, because they knew of no way in which two lenses could be made to do better than one. But every one who has studied optics knows that white light passing through a single lens is not all brought to the same focus, but that the blue light will come to a focus nearer the objective than the red light. There will, in fact, be a succession of images, blue, green, yellow, and red, corresponding to the colors of the spectrum. It is impossible to see these different images clearly at the same time, because each of them will render all the others indistinct.

The achromatic object-glass, invented by Dollond, about 1750, obviates this difficulty, and brings all the rays to nearly the same focus. Nearly every one interested in the subject is aware that this object-glass is composed of two lenses—a concave one of flint-glass and a convex one of crown-glass, the latter being on the side towards the object. This is the one vital part of the telescope, the construction of which involves the greatest difficulty. Once in possession of a perfect object-glass, the rest of the telescope is a matter of little more than constructive skill which there is no difficulty in commanding.

The construction of the object-glass requires two completely distinct processes: the making of the rough glass, which is the work of the glass-maker; and the grinding and polishing into shape, which is the work of the optician. The ordinary glass of commerce will not answer the purpose of the telescope at all, because it is not sufficiently clear and homogeneous. OPTICAL GLASS, as it is called, must be made of materials selected and purified with the greatest care, and worked in a more elaborate manner than is necessary in any other kind of glass. In the time of Dollond it was found scarcely possible to make good disks of flint-glass more than three or four inches in diameter. Early in the present century, Guinand, of Switzerland, invented a process by which disks of much larger size could be produced. In conjunction with the celebrated Fraunhofer he made disks of nine or ten inches in diameter, which were employed by his colaborer in constructing the telescopes which were so famous in their time. He was long supposed to be in possession of some secret method of avoiding the difficulties which his predecessors had met. It is now believed that this secret, if one it was, consisted principally in the constant stirring of the molten glass during the process of manufacture. However this may be, it is a curious historical fact that the most successful makers of these great disks of glass have either been of the family of Guinand, or successors, in the management of the family firm. It was Feil, a son-in-law or near relative, who made the glass from which Clark fabricated the lenses of the great telescope of the Lick Observatory. His successor, Mantois, of Paris, carried the art to a point of perfection never before approached. The transparency and uniformity of his disks as well as the great size to which he was able to carry them would suggest that he and his successors have out-distanced all competitors in the process. He it was who made the great 40-inch lens for the Yerkes Observatory.

As optical glass is now made, the material is constantly stirred with an iron rod during all the time it is melting in the furnace, and after it has begun to cool, until it becomes so stiff that the stirring has to cease. It is then placed, pot and all, in the annealing furnace, where it is kept nearly at a melting heat for three weeks or more, according to the size of the pot. When the furnace has cooled off, the glass is taken out, and the pot is broken from around it, leaving only the central mass of glass. Having such a mass, there is no trouble in breaking it up into pieces of all desirable purity, and sufficiently large for moderate-sized telescopes. But when a great telescope of two feet aperture or upward is to be constructed, very delicate and laborious operations have to be undertaken. The outside of the glass has first to be chipped off, because it is filled with impurities from the material of the pot itself. But this is not all. Veins of unequal density are always found extending through the interior of the mass, no way of avoiding them having yet been discovered. They are supposed to arise from the materials of the pot and stirring rod, which become mixed in with the glass in consequence of the intense heat to which all are subjected. These veins must, so far as possible, be ground or chipped out with the greatest care. The glass is then melted again, pressed into a flat disk, and once more put into the annealing oven. In fact, the operation of annealing must be repeated every time the glass is melted. When cooled, it is again examined for veins, of which great numbers are sure to be found. The problem now is to remove these by cutting and grinding without either breaking the glass in two or cutting a hole through it. If the parts of the glass are once separated, they can never be joined without producing a bad scar at the point of junction. So long, however, as the surface is unbroken, the interior parts of the glass can be changed in form to any extent. Having ground out the veins as far as possible, the glass is to be again melted, and moulded into proper shape. In this mould great care must be taken to have no folding of the surface. Imagining the latter to be a sort of skin enclosing the melted glass inside, it must be raised up wherever the glass is thinnest, and the latter allowed to slowly run together beneath it.

If the disk is of flint, all the veins must be ground out on the first or second trial, because after two or three mouldings the glass will lose its transparency. A crown disk may, however, be melted a number of times without serious injury. In many cases—perhaps the majority—the artisan finds that after all his months of labor he cannot perfectly clear his glass of the noxious veins, and he has to break it up into smaller pieces. When he finally succeeds, the disk has the form of a thin grindstone two feet or upward in diameter, according to the size of the telescope to be made, and from two to three inches in thickness. The glass is then ready for the optician.

The first process to be performed by the optician is to grind the glass into the shape of a lens with perfectly spherical surfaces. The convex surface must be ground in a saucer-shaped tool of corresponding form. It is impossible to make a tool perfectly spherical in the first place, but success may be secured on the geometrical principle that two surfaces cannot fit each other in all positions unless both are perfectly spherical. The tool of the optician is a very simple affair, being nothing more than a plate of iron somewhat larger, perhaps a fourth, than the lens to be ground to the corresponding curvature. In order to insure its changing to fit the glass, it is covered on the interior with a coating of pitch from an eighth to a quarter of an inch thick. This material is admirably adapted to the purpose because it gives way certainly, though very slowly, to the pressure of the glass. In order that it may have room to change its form, grooves are cut through it in both directions, so as to leave it in the form of squares, like those on a chess-board.

It is then sprinkled over with rouge, moistened with water, and gently warmed. The roughly ground lens is then placed upon it, and moved from side to side. The direction of the motion is slightly changed with every stroke, so that after a dozen or so of strokes the lines of motion will lie in every direction on the tool. This change of direction is most readily and easily effected by the operator slowly walking around as he polishes, at the same time the lens is to be slowly turned around either in the opposite direction or more rapidly yet in the same direction, so that the strokes of the polisher shall cross the lens in all directions. This double motion insures every part of the lens coming into contact with every part of the polisher, and moving over it in every direction.

Then whatever parts either of the lens or of the polisher may be too high to form a spherical surface will be gradually worn down, thus securing the perfect sphericity of both.

When the polishing is done by machinery, which is the custom in Europe, with large lenses, the polisher is slid back and forth over the lens by means of a crank attached to a revolving wheel. The polisher is at the same time slowly revolving around a pivot at its centre, which pivot the crank works into, and the glass below it is slowly turned in an opposite direction. Thus the same effect is produced as in the other system. Those who practice this method claim that by thus using machinery the conditions of a uniform polish for every part of the surface can be more perfectly fulfilled than by a hand motion. The results, however, do not support this view. No European optician will claim to do better than the American firm of Alvan Clark and Sons in producing uniformly good object-glasses, and this firm always does the work by hand, moving the glass over the polisher, and not the polisher over the glass.

Having brought both flint and crown glasses into proper figure by this process, they are joined together, and tested by observations either upon a star in the heavens, or some illuminated point at a little distance on the ground. The reflection of the sun from a drop of quicksilver, a thermometer bulb, or even a piece of broken bottle, makes an excellent artificial star. The very best optician will always find that on a first trial his glass is not perfect. He will find that he has not given exactly the proper curves to secure achromatism. He must then change the figure of one or both the glasses by polishing it upon a tool of slightly different curvature. He may also find that there is some spherical aberration outstanding. He must then alter his curve so as to correct this. The correction of these little imperfections in the figures of the lenses so as to secure perfect vision through them is the most difficult branch of the art of the optician, and upon his skill in practising it will depend more than upon anything else his ultimate success and reputation. The shaping of a pair of lenses in the way we have described is not beyond the power of any person of ordinary mechanical ingenuity, possessing the necessary delicacy of touch and appreciation of the problem he is attacking. But to make a perfect objective of considerable size, which shall satisfy all the wants of the astronomer, is an undertaking requiring such accuracy of eyesight, and judgment in determining where the error lies, and such skill in manipulating so as to remove the defects, that the successful men in any one generation can be counted on one's fingers.

In order that the telescope may finally perform satisfactorily it is not sufficient that the lenses should both be of proper figure; they must also both be properly centred in their cells. If either lens is tipped aside, or slid out from its proper central line, the definition will be injured. As this is liable to happen with almost any telescope, we shall explain how the proper adjustment is to be made.

The easiest way to test this adjustment is to set the cell with the two glasses of the objective in it against a wall at night, and going to a short distance, observe the reflection in the glass of the flame of a candle held in the hand. Three or four reflections will be seen from the different surfaces. The observer, holding the candle before his eye, and having his line of sight as close as possible to the flame, must then move until the different images of the flame coincide with each other. If he cannot bring them into coincidence, owing to different pairs coinciding on different sides of the flame, the glasses are not perfectly centred upon each other. When the centring is perfect, the observer having the light in the line of the axes of the lenses, and (if it were possible to do so) looking through the centre of the flame, would see the three or four images all in coincidence. As he cannot see through the flame itself, he must look first on one side and then on the other, and see if the arrangement of the images seen in the lenses is symmetrical. If, going to different distances, he finds no deviation from symmetry, in this respect the adjustment is near enough for all practical purposes.

A more artistic instrument than a simple candle is a small concave reflector pierced through its centre, such as is used by physicians in examining the throat.

Place this reflector in the prolongation of the optical axis, set the candle so that the light from the reflector shall be shown through the glass, and look through the opening. Images of the reflector itself will then be seen in the object-glass, and if the adjustment is perfect, the reflector can be moved so that they will all come into coincidence together.

When the objective is in the tube of the telescope, it is always well to examine this adjustment from time to time, holding the candle so that its light shall shine through the opening perpendicularly upon the object-glass. The observer looks upon one side of the flame, and then upon the other, to see if the images are symmetrical in the different positions. If in order to see them in this way the candle has to be moved to one side of the central line of the tube, the whole objective must be adjusted. If two images coincide in one position of the candle-flame, and two in another position, so that they cannot all be brought together in any position, it shows that the glasses are not properly adjusted in their cell. It may be remarked that this last adjustment is the proper work of the optician, since it is so difficult that the user of the telescope cannot ordinarily effect it. But the perpendicularity of the whole objective to the tube of the telescope is liable to be deranged in use, and every one who uses such an instrument should be able to rectify an error of this kind.

The question may be asked, How much of a telescope can an amateur observer, under any circumstances, make for himself? As a general rule, his work in this direction must be confined to the tube and the mounting. We should not, it is true, dare to assert that any ingenious young man, with a clear appreciation of optical principles, could not soon learn to grind and polish an object-glass for himself by the method we have described, and thus obtain a much better instrument than Galileo ever had at his command. But it would be a wonderful success if his home-made telescope was equal to the most indifferent one which can be bought at an optician's. The objective, complete in itself, can be purchased at prices depending upon the size.

[Footnote: The following is a rough rule for getting an idea of the price of an achromatic objective, made to order, of the finest quality. Take the cube of the diameter in inches, or, which is the same thing, calculate the contents of a cubical box which would hold a sphere of the same diameter as the clear aperture of the glass. The price of the glass will then range from $1 to $1.75 for each cubic inch in this box. For example, the price of a four-inch objective will probably range from $64 to $112. Very small object-glasses of one or two inches may be a little higher than would be given by this rule. Instruments which are not first-class, but will answer most of the purposes of the amateur, are much cheaper.]

The tube for the telescope may be made of paper, by pasting a great number of thicknesses around a long wooden cylinder. A yet better tube is made of a simple wooden box. The best material, however, is metal, because wood and pasteboard are liable both to get out of shape, and to swell under the influence of moisture. Tin, if it be of sufficient thickness, would be a very good material. The brighter it is kept, the better. The work of fitting the objective into one end of a tin tube of double thickness, and properly adjusting it, will probably be quite within the powers of the ordinary amateur. The fitting of the eye-piece into the other end of the tube will require some skill and care both on his own part and that of his tinsmith.

Although the construction of the eye-piece is much easier than that of the objective, since the same accuracy in adjusting the curves is not necessary, yet the price is lower in a yet greater degree, so that the amateur will find it better to buy than to make his eye-piece, unless he is anxious to test his mechanical powers. For a telescope which has no micrometer, the Huyghenian or negative eye-piece, as it is commonly called, is the best. As made by Huyghens, it consists of two plano-convex lenses, with their plane sides next the eye, as shown in the figure.

So far as we have yet described our telescope it is optically complete. If it could be used as a spy-glass by simply holding it in the hand, and pointing at the object we wish to observe, there would be little need of any very elaborate support. But if a telescope, even of the smallest size, is to be used with regularity, a proper "mounting" is as essential as a good instrument. Persons unpractised in the use of such instruments are very apt to underrate the importance of those accessories which merely enable us to point the telescope. An idea of what is wanted in the mounting may readily be formed if the reader will try to look at a star with an ordinary good-sized spy-glass held in the hand, and then imagine the difficulties he meets with multiplied by fifty.

The smaller and cheaper telescopes, as commonly sold, are mounted on a simple little stand, on which the instrument admits of a horizontal and vertical motion. If one only wants to get a few glimpses of a celestial object, this mounting will answer his purpose. But to make anything like a study of a celestial body, the mounting must be an equatorial one; that is, one of the axes around which the telescope moves must be inclined so as to point towards the pole of the heavens, which is near the polar star. This axis will then make an angle with the horizon equal to the latitude of the place. The telescope cannot, however, be mounted directly on this axis, but must be attached to a second one, itself fastened to this one.

When mounted in this way, an object can be followed in its diurnal motion from east to west by turning on the polar axis alone. But if the greatest facility in use is required, this motion must be performed by clock-work. A telescope with this appendage will commonly cost one thousand dollars and upward, so that it is not usually applied to very small ones.

We will now suppose that the reader wishes to purchase a telescope or an object-glass for himself, and to be able to judge of its performance. He must have the object-glass properly adjusted in its tube, and must use the highest power; that is, the smallest eye-piece, which he intends to use in the instrument. Of course he understands that in looking directly at a star or a celestial object it must appear sharp in outline and well defined. But without long practice with good instruments, this will not give him a very definite idea. If the person who selects the telescope is quite unpractised, it is possible that he can make the best test by ascertaining at what distance he can read ordinary print. To do this he should have an eye-piece magnifying about fifty times for each inch of aperture of the telescope. For instance, if his telescope is three inches clear aperture, then his eye-piece should magnify one hundred and fifty times; if the aperture is four inches, one magnifying two hundred times may be used. This magnifying power is, as a general rule, about the highest that can be advantageously used with any telescope. Supposing this magnifying power to be used, this page should be legible at a distance of four feet for every unit of magnifying power of the telescope. For example, with a power of 100, it should be legible at a distance of 400 feet; with a power of 200, at 800 feet, and so on. To put the condition into another shape: if the telescope will read the print at a distance of 150 feet for each inch of aperture with the best magnifying power, its performance is at least not very bad. If the magnifying power is less than would be given by this rule, the telescope should perform a little better; for instance, a three-inch telescope with a power of 60 should make this page legible at a distance of 300 feet, or four feet for each unit of power.

The test applied by the optician is much more exact, and also more easy. He points the instrument at a star, or at the reflection of the sun's rays from a small round piece of glass or a globule of quicksilver several hundred yards away, and ascertains whether the rays are all brought to a focus. This is not done by simply looking at the star, but by alternately pushing the eye-piece in beyond the point of distinct vision and drawing it out past the point. In this way the image of the star will appear, not as a point, but as a round disk of light. If the telescope is perfect, this disk will appear round and of uniform brightness in either position of the eye-piece. But if there is any spherical aberration or differences of density in different parts of the glass, the image will appear distorted in various ways. If the spherical aberration is not correct, the outer rim of the disk will be brighter than the centre when the eye-piece is pushed in, and the centre will be the brighter when it is drawn out. If the curves of the glass are not even all around, the image will appear oval in one or the other position. If there are large veins of unequal density, wings or notches will be seen on the image. If the atmosphere is steady, the image, when the eye-piece is pushed in, will be formed of a great number of minute rings of light. If the glass is good, these rings will be round, unbroken, and equally bright. We present several figures showing how these spectral images, as they are sometimes called, will appear; first, when the eye-piece is pushed in, and secondly, when it is drawn out, with telescopes of different qualities.

We have thus far spoken only of the refracting telescope, because it is the kind with which an observer would naturally seek to supply himself. At the same time there is little doubt that the construction of a reflector of moderate size is easier than that of a corresponding refractor. The essential part of the reflector is a slightly concave mirror of any metal which will bear a high polish. This mirror may be ground and polished in the same way as a lens, only the tool must be convex.

A The telescope is all right B Spherical aberration shown by the light and dark centre C The objective is not spherical but elliptical D The glass not uniform—a very bad and incurable case E One side of the objective nearer than the other. Adjust it]

Of late years it has become very common to make the mirror of glass and to cover the reflecting face with an exceedingly thin film of silver, which can be polished by hand in a few minutes. Such a mirror differs from our ordinary looking-glass in that the coating of silver is put on the front surface, so that the light does not pass through the glass. Moreover, the coating of silver is so thin as to be almost transparent: in fact, the sun may be seen through it by direct vision as a faint blue object. Silvered glass reflectors made in this way are extensively manufactured in London, and are far cheaper than refracting telescopes of corresponding size. Their great drawback is the want of permanence in the silver film. In the city the film will ordinarily tarnish in a few months from the sulphurous vapors arising from gaslights and other sources, and even in the country it is very difficult to preserve the mirror from the contact of everything that will injure it. In consequence, the possessor of such a telescope, if he wishes to keep it in order, must always be prepared to resilver and repolish it. To do this requires such careful manipulation and management of the chemicals that it is hardly to be expected that an amateur will take the trouble to keep his telescope in order, unless he has a taste for chemistry as well as for astronomy.

The curiosity to see the heavenly bodies through great telescopes is so wide-spread that we are apt to forget how much can be seen and done with small ones. The fact is that a large proportion of the astronomical observations of past times have been made with what we should now regard as very small instruments, and a good deal of the solid astronomical work of the present time is done with meridian circles the apertures of which ordinarily range from four to eight inches. One of the most conspicuous examples in recent times of how a moderate-sized instrument may be utilized is afforded by the discoveries of double stars made by Mr. S. W. Burnham, of Chicago. Provided with a little six-inch telescope, procured at his own expense from the Messrs. Clark, he has discovered many hundred double stars so difficult that they had escaped the scrutiny of Maedler and the Struves, and gained for himself one of the highest positions among the astronomers of the day engaged in the observation of these objects. It was with this little instrument that on Mount Hamilton, California—afterward the site of the great Lick Observatory—he discovered forty-eight new double stars, which had remained unnoticed by all previous observers. First among the objects which show beautifully through moderate instruments stands the moon. People who want to see the moon at an observatory generally make the mistake of looking when the moon is full, and asking to see it through the largest telescope. Nothing can then be made out but a brilliant blaze of light, mottled with dark spots, and crossed by irregular bright lines. The best time to view the moon is near or before the first quarter, or when she is from three to eight days old. The last quarter is of course equally favorable, so far as seeing is concerned, only one must be up after midnight to see her in that position. Seen through a three or four inch telescope, a day or two before the first quarter, about half an hour after sunset, and with a magnifying power between fifty and one hundred, the moon is one of the most beautiful objects in the heavens. Twilight softens her radiance so that the eye is not dazzled as it will be when the sky is entirely dark. The general aspect she then presents is that of a hemisphere of beautiful chased silver carved out in curious round patterns with a more than human skill. If, however, one wishes to see the minute details of the lunar surface, in which many of our astronomers are now so deeply interested, he must use a higher magnifying power. The general beautiful effect is then lessened, but more details are seen. Still, it is hardly necessary to seek for a very large telescope for any investigation of the lunar surface. I very much doubt whether any one has ever seen anything on the moon which could not be made out in a clear, steady atmosphere with a six-inch telescope of the first class.

Next to the moon, Saturn is among the most beautiful of celestial objects. Its aspect, however, varies with its position in its orbit. Twice in the course of a revolution, which occupies nearly thirty years, the rings are seen edgewise, and for a few days are invisible even in a powerful telescope. For an entire year their form may be difficult to make out with a small telescope. These unfavorable conditions occur in 1907 and 1921. Between these dates, especially for some years after 1910, the position of the planet in the sky will be the most favorable, being in northern declination, near its perihelion, and having its rings widely open. We all know that Saturn is plainly visible to the naked eye, shining almost like a star of the first magnitude, so that there is no difficulty in finding it if one knows when and where to look. In 1906-1908 its oppositions occur in the month of September. In subsequent years, it will occur a month later every two and a half years. The ring can be seen with a common, good spy-glass fastened to a post so as to be steady. A four or five-inch telescope will show most of the satellites, the division in the ring, and, when the ring is well opened, the curious dusky ring discovered by Bond. This "crape ring," as it is commonly called, is one of the most singular phenomena presented by that planet.

It might be interesting to the amateur astronomer with a keen eye and a telescope of four inches aperture or upward to frequently scrutinize Saturn, with a view of detecting any extraordinary eruptions upon his surface, like that seen by Professor Hall in 1876. On December 7th of that year a bright spot was seen upon Saturn's equator. It elongated itself from day to day, and remained visible for several weeks. Such a thing had never before been known upon this planet, and had it not been that Professor Hall was engaged in observations upon the satellites, it would not have been seen then. A similar spot on the planet was recorded in 1902, and much more extensively noticed. On this occasion the spot appeared in a higher latitude from the planet's equator than did Professor Hall's. At this appearance the time of the planet's revolution on its axis was found to be somewhat greater than in 1876, in accordance with the general law exhibited in the rotations of the sun and of Jupiter. Notwithstanding their transient character, these two spots have afforded the only determination of the time of revolution of Saturn which has been made since Herschel the elder.

Of the satellites of Saturn the brightest is Titan, which can be seen with the smallest telescope, and revolves around the planet in fifteen days. Iapetus, the outer satellite, is remarkable for varying greatly in brilliancy during its revolution around the planet. Any one having the means and ability to make accurate photometrical estimates of the light of this satellite in all points of its orbit, can thereby render a valuable service to astronomy.

The observations of Venus, by which the astronomers of the last century supposed themselves to have discovered its time of rotation on its axis, were made with telescopes much inferior to ours. Although their observations have not been confirmed, some astronomers are still inclined to think that their results have not been refuted by the failure of recent observers to detect those changes which the older ones describe on the surface of the planet. With a six-inch telescope of the best quality, and with time to choose the most favorable moment, one will be as well equipped to settle the question of the rotation of Venus as the best observer. The few days near each inferior conjunction are especially to be taken advantage of. The questions to be settled are two: first, are there any dark spots or other markings on the disk? second, are there any irregularities in the form of the sharp cusps? The central portions of the disk are much darker than the outline, and it is probably this fact which has given rise to the impression of dark spots. Unless this apparent darkness changes from time to time, or shows some irregularity in its outline, it cannot indicate any rotation of the planet. The best time to scrutinize the sharp cusps will be when the planet is nearly on the line from the earth to the sun. The best hour of the day is near sunset, the half-hour following sunset being the best of all. But if Venus is near the sun, she will after sunset be too low down to be well seen, and must be looked at late in the afternoon.

The planet Mars must always be an object of great interest, because of all the heavenly bodies it is that which appears to bear the greatest resemblance to the earth. It comes into opposition at intervals of a little more than two years, and can be well seen only for a month or two before and after each opposition. It is hopeless to look for the satellites of Mars with any but the greatest telescopes of the world. But the markings on the surface, from which the time of rotation has been determined, and which indicate a resemblance to the surface of our own planet, can be well seen with telescopes of six inches aperture and upward. One or both of the bright polar spots, which are supposed to be due to deposits of snow, can be seen with smaller telescopes when the situation of the planet is favorable.

The case is different with the so-called canals discovered by Schiaparelli in 1877, which have ever since excited so much interest, and given rise to so much discussion as to their nature. The astronomer who has had the best opportunities for studying them is Mr. Percival Lowell, whose observatory at Flaggstaff, Arizona, is finely situated for the purpose, while he also has one of the best if not the largest of telescopes. There the canals are seen as fine dark lines; but, even then, they must be fifty miles in breadth, so that the word "canal" may be regarded as a misnomer.

Although the planet Jupiter does not present such striking features as Saturn, it is of even more interest to the amateur astronomer, because he can study it with less optical power, and see more of the changes upon its surface. Every work on astronomy tells in a general way of the belts of Jupiter, and many speculate upon their causes. The reader of recent works knows that Jupiter is supposed to be not a solid mass like the earth, but a great globe of molten and vaporous matter, intermediate in constitution between the earth and the sun. The outer surface which we see is probably a hot mass of vapor hundreds of miles deep, thrown up from the heated interior. The belts are probably cloudlike forms in this vaporous mass.

Certain it is that they are continually changing, so that the planet seldom looks exactly the same on two successive evenings. The rotation of the planet can be very well seen by an hour's watching. In two hours an object at the centre of the disk will move off to near the margin.

The satellites of this planet, in their ever-varying phases, are objects of perennial interest. Their eclipses may be observed with a very small telescope, if one knows when to look for them. To do this successfully, and without waste of time, it is necessary to have an astronomical ephemeris for the year. All the observable phenomena are there predicted for the convenience of observers. Perhaps the most curious observation to be made is that of the shadow of the satellite crossing the disk of Jupiter. The writer has seen this perfectly with a six-inch telescope, and a much smaller one would probably show it well. With a telescope of this size, or a little larger, the satellites can be seen between us and Jupiter. Sometimes they appear a little brighter than the planet, and sometimes a little fainter.

Of the remaining large planets, Mercury, the inner one, and Uranus and Neptune, the two outer ones, are of less interest than the others to an amateur with a small telescope, because they are more difficult to see. Mercury can, indeed, be observed with the smallest instrument, but no physical configurations or changes have ever been made out upon his surface. The question whether any such can be observed is still an open one, which can be settled only by long and careful scrutiny. A small telescope is almost as good for this purpose as a large one, because the atmospheric difficulties in the way of getting a good view of the planet cannot be lessened by an increase of telescopic power.

Uranus and Neptune are so distant that telescopes of considerable size and high magnifying power are necessary to show their disks. In small telescopes they have the appearance of stars, and the observer has no way of distinguishing them from the surrounding stars unless he can command the best astronomical appliances, such as star maps, circles on his instrument, etc. It is, however, to be remarked, as a fact not generally known, that Uranus can be well seen with the naked eye if one knows where to look for it. To recognize it, it is necessary to have an astronomical ephemeris showing its right ascension and declination, and star maps showing where the parallels of right ascension and declination lie among the stars. When once found by the naked eye, there will, of course, be no difficulty in pointing the telescope upon it.

Of celestial objects which it is well to keep a watch upon, and which can be seen to good advantage with inexpensive instruments, the sun may be considered as holding the first place. Astronomers who make a specialty of solar physics have, especially in this country, so many other duties, and their view is so often interrupted by clouds, that a continuous record of the spots on the sun and the changes they undergo is hardly possible. Perhaps one of the most interesting and useful pieces of astronomical work which an amateur can perform will consist of a record of the origin and changes of form of the solar spots and faculae. What does a spot look like when it first comes into sight? Does it immediately burst forth with considerable magnitude, or does it begin as the smallest visible speck, and gradually grow? When several spots coalesce into one, how do they do it? When a spot breaks up into several pieces, what is the seeming nature of the process? How do the groups of brilliant points called faculae come, change, and grow? All these questions must no doubt be answered in various ways, according to the behavior of the particular spot, but the record is rather meagre, and the conscientious and industrious amateur will be able to amuse himself by adding to it, and possibly may make valuable contributions to science in the same way.

Still another branch of astronomical observation, in which industry and skill count for more than expensive instruments, is the search for new comets. This requires a very practised eye, in order that the comet may be caught among the crowd of stars which flit across the field of view as the telescope is moved. It is also necessary to be well acquainted with a number of nebulae which look very much like comets. The search can be made with almost any small telescope, if one is careful to use a very low power. With a four-inch telescope a power not exceeding twenty should be employed. To search with ease, and in the best manner, the observer should have what among astronomers is familiarly known as a "broken-backed telescope." This instrument has the eye-piece on the end of the axis, where one would never think of looking for it. By turning the instrument on this axis, it sweeps from one horizon through the zenith and over to the other horizon without the observer having to move his head. This is effected by having a reflector in the central part of the instrument, which throws the rays of light at right angles through the axis.

How well this search can be conducted by observers with limited means at their disposal is shown by the success of several American observers, among whom Messrs. W. R. Brooks, E. E. Barnard, and Lewis Swift are well known. The cometary discoveries of these men afford an excellent illustration of how much can be done with the smallest means when one sets to work in the right spirit.

The larger number of wonderful telescopic objects are to be sought for far beyond the confines of the solar system, in regions from which light requires years to reach us. On account of their great distance, these objects generally require the most powerful telescopes to be seen in the best manner; but there are quite a number within the range of the amateur. Looking at the Milky Way, especially its southern part, on a clear winter or summer evening, tufts of light will be seen here and there. On examining these tufts with a telescope, they will be found to consist of congeries of stars. Many of these groups are of the greatest beauty, with only a moderate optical power. Of all the groups in the Milky Way the best known is that in the sword-handle of Perseus, which may be seen during the greater part of the year, and is distinctly visible to the naked eye as a patch of diffused light. With the telescope there are seen in this patch two closely connected clusters of stars, or perhaps we ought rather to say two centres of condensation.

Another object of the same class is Proesepe in the constellation Cancer. This can be very distinctly seen by the naked eye on a clear moonless night in winter or spring as a faint nebulous object, surrounded by three small stars. The smallest telescope shows it as a group of stars.

Of all stellar objects, the great nebula of Orion is that which has most fascinated the astronomers of two centuries. It is distinctly visible to the naked eye, and may be found without difficulty on any winter night. The three bright stars forming the sword-belt of Orion are known to every one who has noticed that constellation. Below this belt is seen another triplet of stars, not so bright, and lying in a north and south direction. The middle star of this triplet is the great nebula. At first the naked eye sees nothing to distinguish it from other stars, but if closely scanned it will be seen to have a hazy aspect. A four-inch telescope will show its curious form. Not the least interesting of its features are the four stars known as the "Trapezium," which are located in a dark region near its centre. In fact, the whole nebula is dotted with stars, which add greatly to the effect produced by its mysterious aspect.

The great nebula of Andromeda is second only to that of Orion in interest. Like the former, it is distinctly visible to the naked eye, having the aspect of a faint comet. The most curious feature of this object is that although the most powerful telescopes do not resolve it into stars, it appears in the spectroscope as if it were solid matter shining by its own light.

The above are merely selections from the countless number of objects which the heavens offer to telescopic study. Many such are described in astronomical works, but the amateur can gratify his curiosity to almost any extent by searching them out for himself.

Ever since 1878 a red spot, unlike any before noticed, has generally been visible on Jupiter. At first it was for several years a very conspicuous object, but gradually faded away, so that since 1890 it has been made out only with difficulty. But it is now regarded as a permanent feature of the planet. There is some reason to believe it was occasionally seen long before attention was first attracted to it. Doubtless, when it can be seen at all, practice in observing such objects is more important than size of telescope.

WHAT THE ASTRONOMERS ARE DOING

In no field of science has human knowledge been more extended in our time than in that of astronomy. Forty years ago astronomical research seemed quite barren of results of great interest or value to our race. The observers of the world were working on a traditional system, grinding out results in an endless course, without seeing any prospect of the great generalizations to which they might ultimately lead. Now this is all changed. A new instrument, the spectroscope, has been developed, the extent of whose revelations we are just beginning to learn, although it has been more than thirty years in use. The application of photography has been so extended that, in some important branches of astronomical work, the observer simply photographs the phenomenon which he is to study, and then makes his observation on the developed negative.

The world of astronomy is one of the busiest that can be found to-day, and the writer proposes, with the reader's courteous consent, to take him on a stroll through it and see what is going on. We may begin our inspection with a body which is, for us, next to the earth, the most important in the universe. I mean the sun. At the Greenwich Observatory the sun has for more than twenty years been regularly photographed on every clear day, with the view of determining the changes going on in its spots. In recent years these observations have been supplemented by others, made at stations in India and Mauritius, so that by the combination of all it is quite exceptional to have an entire day pass without at least one photograph being taken. On these observations must mainly rest our knowledge of the curious cycle of change in the solar spots, which goes through a period of about eleven years, but of which no one has as yet been able to establish the cause.

This Greenwich system has been extended and improved by an American.

Professor George E. Hale, formerly Director of the Yerkes Observatory, has devised an instrument for taking photographs of the sun by a single ray of the spectrum. The light emitted by calcium, the base of lime, and one of the substances most abundant in the sun, is often selected to impress the plate.

The Carnegie Institution has recently organized an enterprise for carrying on the study of the sun under a combination of better conditions than were ever before enjoyed. The first requirement in such a case is the ablest and most enthusiastic worker in the field, ready to devote all his energies to its cultivation. This requirement is found in the person of Professor Hale himself. The next requirement is an atmosphere of the greatest transparency, and a situation at a high elevation above sea-level, so that the passage of light from the sun to the observer shall be obstructed as little as possible by the mists and vapors near the earth's surface. This requirement is reached by placing the observatory on Mount Wilson, near Pasadena, California, where the climate is found to be the best of any in the United States, and probably not exceeded by that of any other attainable point in the world. The third requirement is the best of instruments, specially devised to meet the requirements. In this respect we may be sure that nothing attainable by human ingenuity will be found wanting.

Thus provided, Professor Hale has entered upon the task of studying the sun, and recording from day to day all the changes going on in it, using specially devised instruments for each purpose in view. Photography is made use of through almost the entire investigation. A full description of the work would require an enumeration of technical details, into which we need not enter at present. Let it, therefore, suffice to say in a general way that the study of the sun is being carried on on a scale, and with an energy worthy of the most important subject that presents itself to the astronomer. Closely associated with this work is that of Professor Langley and Dr. Abbot, at the Astro-Physical Observatory of the Smithsonian Institution, who have recently completed one of the most important works ever carried out on the light of the sun. They have for years been analyzing those of its rays which, although entirely invisible to our eyes, are of the same nature as those of light, and are felt by us as heat. To do this, Langley invented a sort of artificial eye, which he called a bolometer, in which the optic nerve is made of an extremely thin strip of metal, so slight that one can hardly see it, which is traversed by an electric current. This eye would be so dazzled by the heat radiated from one's body that, when in use, it must be protected from all such heat by being enclosed in a case kept at a constant temperature by being immersed in water. With this eye the two observers have mapped the heat rays of the sun down to an extent and with a precision which were before entirely unknown.

The question of possible changes in the sun's radiation, and of the relation of those changes to human welfare, still eludes our scrutiny. With all the efforts that have been made, the physicist of to-day has not yet been able to make anything like an exact determination of the total amount of heat received from the sun. The largest measurements are almost double the smallest. This is partly due to the atmosphere absorbing an unknown and variable fraction of the sun's rays which pass through it, and partly to the difficulty of distinguishing the heat radiated by the sun from that radiated by terrestrial objects.

In one recent instance, a change in the sun's radiation has been noticed in various parts of the world, and is of especial interest because there seems to be little doubt as to its origin. In the latter part of 1902 an extraordinary diminution was found in the intensity of the sun's heat, as measured by the bolometer and other instruments. This continued through the first part of 1903, with wide variations at different places, and it was more than a year after the first diminution before the sun's rays again assumed their ordinary intensity.

This result is now attributed to the eruption of Mount Pelee, during which an enormous mass of volcanic dust and vapor was projected into the higher regions of the air, and gradually carried over the entire earth by winds and currents. Many of our readers may remember that something yet more striking occurred after the great cataclasm at Krakatoa in 1883, when, for more than a year, red sunsets and red twilights of a depth of shade never before observed were seen in every part of the world.

What we call universology—the knowledge of the structure and extent of the universe—must begin with a study of the starry heavens as we see them. There are perhaps one hundred million stars in the sky within the reach of telescopic vision. This number is too great to allow of all the stars being studied individually; yet, to form the basis for any conclusion, we must know the positions and arrangement of as many of them as we can determine.

To do this the first want is a catalogue giving very precise positions of as many of the brighter stars as possible. The principal national observatories, as well as some others, are engaged in supplying this want. Up to the present time about 200,000 stars visible in our latitudes have been catalogued on this precise plan, and the work is still going on. In that part of the sky which we never see, because it is only visible from the southern hemisphere, the corresponding work is far from being as extensive. Sir David Gill, astronomer at the Cape of Good Hope, and also the directors of other southern observatories, are engaged in pushing it forward as rapidly as the limited facilities at their disposal will allow.

Next in order comes the work of simply listing as many stars as possible. Here the most exact positions are not required. It is only necessary to lay down the position of each star with sufficient exactness to distinguish it from all its neighbors. About 400,000 stars were during the last half-century listed in this way at the observatory of Bonn by Argelander, Schonfeld, and their assistants. This work is now being carried through the southern hemisphere on a large scale by Thome, Director of the Cordoba Observatory, in the Argentine Republic. This was founded thirty years ago by our Dr. B. A. Gould, who turned it over to Dr. Thome in 1886. The latter has, up to the present time, fixed and published the positions of nearly half a million stars. This work of Thome extends to fainter stars than any other yet attempted, so that, as it goes on, we have more stars listed in a region invisible in middle northern latitudes than we have for that part of the sky we can see. Up to the present time three quarto volumes giving the positions and magnitudes of the stars have appeared. Two or three volumes more, and, perhaps, ten or fifteen years, will be required to complete the work.

About twenty years ago it was discovered that, by means of a telescope especially adapted to this purpose, it was possible to photograph many more stars than an instrument of the same size would show to the eye. This discovery was soon applied in various quarters. Sir David Gill, with characteristic energy, photographed the stars of the southern sky to the number of nearly half a million. As it was beyond his power to measure off and compute the positions of the stars from his plates, the latter were sent to Professor J. C. Kapteyn, of Holland, who undertook the enormous labor of collecting them into a catalogue, the last volume of which was published in 1899. One curious result of this enterprise is that the work of listing the stars is more complete for the southern hemisphere than for the northern.

Another great photographic work now in progress has to do with the millions of stars which it is impossible to handle individually. Fifteen years ago an association of observatories in both hemispheres undertook to make a photographic chart of the sky on the largest scale. Some portions of this work are now approaching completion, but in others it is still in a backward state, owing to the failure of several South American observatories to carry out their part of the programme. When it is all done we shall have a picture of the sky, the study of which may require the labor of a whole generation of astronomers.

Quite independently of this work, the Harvard University, under the direction of Professor Pickering, keeps up the work of photographing the sky on a surprising scale. On this plan we do not have to leave it to posterity to learn whether there is any change in the heavens, for one result of the enterprise has been the discovery of thirteen of the new stars which now and then blaze out in the heavens at points where none were before known. Professor Pickering's work has been continually enlarged and improved until about 150,000 photographic plates, showing from time to time the places of countless millions of stars among their fellows are now stored at the Harvard Observatory. Not less remarkable than this wealth of material has been the development of skill in working it up. Some idea of the work will be obtained by reflecting that, thirty years ago, careful study of the heavens by astronomers devoting their lives to the task had resulted in the discovery of some two or three hundred stars, varying in their light. Now, at Harvard, through keen eyes studying and comparing successive photographs not only of isolated stars, but of clusters and agglomerations of stars in the Milky Way and elsewhere, discoveries of such objects numbering hundreds have been made, and the work is going on with ever-increasing speed. Indeed, the number of variable stars now known is such that their study as individual objects no longer suffices, and they must hereafter be treated statistically with reference to their distribution in space, and their relations to one another, as a census classifies the entire population without taking any account of individuals.

The works just mentioned are concerned with the stars. But the heavenly spaces contain nebulae as well as stars; and photography can now be even more successful in picturing them than the stars. A few years ago the late lamented Keeler, at the Lick Observatory, undertook to see what could be done by pointing the Crossley reflecting telescope at the sky and putting a sensitive photographic plate in the focus. He was surprised to find that a great number of nebulae, the existence of which had never before been suspected, were impressed on the plate. Up to the present time the positions of about 8000 of these objects have been listed. Keeler found that there were probably 200,000 nebulae in the heavens capable of being photographed with the Crossley reflector. But the work of taking these photographs is so great, and the number of reflecting telescopes which can be applied to it so small, that no one has ventured to seriously commence it. It is worthy of remark that only a very small fraction of these objects which can be photographed are visible to the eye, even with the most powerful telescope.

This demonstration of what the reflecting telescope can do may be regarded as one of the most important discoveries of our time as to the capabilities of astronomical instruments. It has long been known that the image formed in the focus of the best refracting telescope is affected by an imperfection arising from the different action of the glasses on rays of light of different colors. Hence, the image of a star can never be seen or photographed with such an instrument, as an actual point, but only as a small, diffused mass. This difficulty is avoided in the reflecting telescope; but a new difficulty is found in the bending of the mirror under the influence of its own weight. Devices for overcoming this had been so far from successful that, when Mr. Crossley presented his instrument to the Lick Observatory, it was feared that little of importance could be done with it. But as often happens in human affairs outside the field of astronomy, when ingenious and able men devote their attention to the careful study of a problem, it was found that new results could be reached. Thus it was that, before a great while, what was supposed to be an inferior instrument proved not only to have qualities not before suspected, but to be the means of making an important addition to the methods of astronomical investigation.

In order that our knowledge of the position of a star may be complete, we must know its distance. This can be measured only through the star's parallax—that is to say, the slight change in its direction produced by the swing of our earth around its orbit. But so vast is the distance in question that this change is immeasurably small, except for, perhaps, a few hundred stars, and even for these few its measurement almost baffles the skill of the most expert astronomer. Progress in this direction is therefore very slow, and there are probably not yet a hundred stars of which the parallax has been ascertained with any approach to certainty. Dr. Chase is now completing an important work of this kind at the Yale Observatory.

To the most refined telescopic observations, as well as to the naked eye, the stars seem all alike, except that they differ greatly in brightness, and somewhat in color. But when their light is analyzed by the spectroscope, it is found that scarcely any two are exactly alike. An important part of the work of the astro-physical observatories, especially that of Harvard, consists in photographing the spectra of thousands of stars, and studying the peculiarities thus brought out. At Harvard a large portion of this work is done as part of the work of the Henry Draper Memorial, established by his widow in memory of the eminent investigator of New York, who died twenty years ago.

By a comparison of the spectra of stars Sir William Huggins has developed the idea that these bodies, like human beings, have a life history. They are nebulae in infancy, while the progress to old age is marked by a constant increase in the density of their substance. Their temperature also changes in a way analogous to the vigor of the human being. During a certain time the star continually grows hotter and hotter. But an end to this must come, and it cools off in old age. What the age of a star may be is hard even to guess. It is many millions of years, perhaps hundreds, possibly even thousands, of millions.

Some attempt at giving the magnitude is included in every considerable list of stars. The work of determining the magnitudes with the greatest precision is so laborious that it must go on rather slowly. It is being pursued on a large scale at the Harvard Observatory, as well as in that of Potsdam, Germany.

We come now to the question of changes in the appearance of bright stars. It seems pretty certain that more than one per cent of these bodies fluctuate to a greater or less extent in their light. Observations of these fluctuations, in the case of at least the brighter stars, may be carried on without any instrument more expensive than a good opera-glass—in fact, in the case of stars visible to the naked eye, with no instrument at all.

As a general rule, the light of these stars goes through its changes in a regular period, which is sometimes as short as a few hours, but generally several days, frequently a large fraction of a year or even eighteen months. Observations of these stars are made to determine the length of the period and the law of variation of the brightness. Any person with a good eye and skill in making estimates can make the observations if he will devote sufficient pains to training himself; but they require a degree of care and assiduity which is not to be expected of any one but an enthusiast on the subject. One of the most successful observers of the present time is Mr. W. A. Roberts, a resident of South Africa, whom the Boer war did not prevent from keeping up a watch of the southern sky, which has resulted in greatly increasing our knowledge of variable stars. There are also quite a number of astronomers in Europe and America who make this particular study their specialty.

During the past fifteen years the art of measuring the speed with which a star is approaching us or receding from us has been brought to a wonderful degree of perfection. The instrument with which this was first done was the spectroscope; it is now replaced with another of the same general kind, called the spectrograph. The latter differs from the other only in that the spectrum of the star is photographed, and the observer makes his measures on the negative. This method was first extensively applied at the Potsdam Observatory in Germany, and has lately become one of the specialties of the Lick Observatory, where Professor Campbell has brought it to its present degree of perfection. The Yerkes Observatory is also beginning work in the same line, where Professor Frost is already rivalling the Lick Observatory in the precision of his measures.

Let us now go back to our own little colony and see what is being done to advance our knowledge of the solar system. This consists of planets, on one of which we dwell, moons revolving around them, comets, and meteoric bodies. The principal national observatories keep up a more or less orderly system of observations of the positions of the planets and their satellites in order to determine the laws of their motion. As in the case of the stars, it is necessary to continue these observations through long periods of time in order that everything possible to learn may be discovered.

Our own moon is one of the enigmas of the mathematical astronomer. Observations show that she is deviating from her predicted place, and that this deviation continues to increase. True, it is not very great when measured by an ordinary standard. The time at which the moon's shadow passed a given point near Norfolk during the total eclipse of May 29, 1900, was only about seven seconds different from the time given in the Astronomical Ephemeris. The path of the shadow along the earth was not out of place by more than one or two miles But, small though these deviations are, they show that something is wrong, and no one has as yet found out what it is. Worse yet, the deviation is increasing rapidly. The observers of the total eclipse in August, 1905, were surprised to find that it began twenty seconds before the predicted time. The mathematical problems involved in correcting this error are of such complexity that it is only now and then that a mathematician turns up anywhere in the world who is both able and bold enough to attack them.

There now seems little doubt that Jupiter is a miniature sun, only not hot enough at its surface to shine by its own light The point in which it most resembles the sun is that its equatorial regions rotate in less time than do the regions near the poles. This shows that what we see is not a solid body. But none of the careful observers have yet succeeded in determining the law of this difference of rotation.

Twelve years ago a suspicion which had long been entertained that the earth's axis of rotation varied a little from time to time was verified by Chandler. The result of this is a slight change in the latitude of all places on the earth's surface, which admits of being determined by precise observations. The National Geodetic Association has established four observatories on the same parallel of latitude—one at Gaithersburg, Maryland, another on the Pacific coast, a third in Japan, and a fourth in Italy—to study these variations by continuous observations from night to night. This work is now going forward on a well-devised plan.

A fact which will appeal to our readers on this side of the Atlantic is the success of American astronomers. Sixty years ago it could not be said that there was a well-known observatory on the American continent. The cultivation of astronomy was confined to a professor here and there, who seldom had anything better than a little telescope with which he showed the heavenly bodies to his students. But during the past thirty years all this has been changed. The total quantity of published research is still less among us than on the continent of Europe, but the number of men who have reached the highest success among us may be judged by one fact. The Royal Astronomical Society of England awards an annual medal to the English or foreign astronomer deemed most worthy of it. The number of these medals awarded to Americans within twenty-five years is about equal to the number awarded to the astronomers of all other nations foreign to the English. That this preponderance is not growing less is shown by the award of medals to Americans in three consecutive years—1904, 1905, and 1906. The recipients were Hale, Boss, and Campbell. Of the fifty foreign associates chosen by this society for their eminence in astronomical research, no less than eighteen—more than one-third—are Americans.

LIFE IN THE UNIVERSE

So far as we can judge from what we see on our globe, the production of life is one of the greatest and most incessant purposes of nature. Life is absent only in regions of perpetual frost, where it never has an opportunity to begin; in places where the temperature is near the boiling-point, which is found to be destructive to it; and beneath the earth's surface, where none of the changes essential to it can come about. Within the limits imposed by these prohibitory conditions—that is to say, within the range of temperature at which water retains its liquid state, and in regions where the sun's rays can penetrate and where wind can blow and water exist in a liquid form—life is the universal rule. How prodigal nature seems to be in its production is too trite a fact to be dwelt upon. We have all read of the millions of germs which are destroyed for every one that comes to maturity. Even the higher forms of life are found almost everywhere. Only small islands have ever been discovered which were uninhabited, and animals of a higher grade are as widely diffused as man.

If it would be going too far to claim that all conditions may have forms of life appropriate to them, it would be going as much too far in the other direction to claim that life can exist only with the precise surroundings which nurture it on this planet. It is very remarkable in this connection that while in one direction we see life coming to an end, in the other direction we see it flourishing more and more up to the limit. These two directions are those of heat and cold. We cannot suppose that life would develop in any important degree in a region of perpetual frost, such as the polar regions of our globe. But we do not find any end to it as the climate becomes warmer. On the contrary, every one knows that the tropics are the most fertile regions of the globe in its production. The luxuriance of the vegetation and the number of the animals continually increase the more tropical the climate becomes. Where the limit may be set no one can say. But it would doubtless be far above the present temperature of the equatorial regions.

It has often been said that this does not apply to the human race, that men lack vigor in the tropics. But human vigor depends on so many conditions, hereditary and otherwise, that we cannot regard the inferior development of humanity in the tropics as due solely to temperature. Physically considered, no men attain a better development than many tribes who inhabit the warmer regions of the globe. The inferiority of the inhabitants of these regions in intellectual power is more likely the result of race heredity than of temperature.

We all know that this earth on which we dwell is only one of countless millions of globes scattered through the wilds of infinite space. So far as we know, most of these globes are wholly unlike the earth, being at a temperature so high that, like our sun, they shine by their own light. In such worlds we may regard it as quite certain that no organized life could exist. But evidence is continually increasing that dark and opaque worlds like ours exist and revolve around their suns, as the earth on which we dwell revolves around its central luminary. Although the number of such globes yet discovered is not great, the circumstances under which they are found lead us to believe that the actual number may be as great as that of the visible stars which stud the sky. If so, the probabilities are that millions of them are essentially similar to our own globe. Have we any reason to believe that life exists on these other worlds?

The reader will not expect me to answer this question positively. It must be admitted that, scientifically, we have no light upon the question, and therefore no positive grounds for reaching a conclusion. We can only reason by analogy and by what we know of the origin and conditions of life around us, and assume that the same agencies which are at play here would be found at play under similar conditions in other parts of the universe.

If we ask what the opinion of men has been, we know historically that our race has, in all periods of its history, peopled other regions with beings even higher in the scale of development than we are ourselves. The gods and demons of an earlier age all wielded powers greater than those granted to man—powers which they could use to determine human destiny. But, up to the time that Copernicus showed that the planets were other worlds, the location of these imaginary beings was rather indefinite. It was therefore quite natural that when the moon and planets were found to be dark globes of a size comparable with that of the earth itself, they were made the habitations of beings like unto ourselves.

The trend of modern discovery has been against carrying this view to its extreme, as will be presently shown. Before considering the difficulties in the way of accepting it to the widest extent, let us enter upon some preliminary considerations as to the origin and prevalence of life, so far as we have any sound basis to go upon.

A generation ago the origin of life upon our planet was one of the great mysteries of science. All the facts brought out by investigation into the past history of our earth seemed to show, with hardly the possibility of a doubt, that there was a time when it was a fiery mass, no more capable of serving as the abode of a living being than the interior of a Bessemer steel furnace. There must therefore have been, within a certain period, a beginning of life upon its surface. But, so far as investigation had gone—indeed, so far as it has gone to the present time—no life has been found to originate of itself. The living germ seems to be necessary to the beginning of any living form. Whence, then, came the first germ? Many of our readers may remember a suggestion by Sir William Thomson, now Lord Kelvin, made twenty or thirty years ago, that life may have been brought to our planet by the falling of a meteor from space. This does not, however, solve the difficulty—indeed, it would only make it greater. It still leaves open the question how life began on the meteor; and granting this, why it was not destroyed by the heat generated as the meteor passed through the air. The popular view that life began through a special act of creative power seemed to be almost forced upon man by the failure of science to discover any other beginning for it. It cannot be said that even to-day anything definite has been actually discovered to refute this view. All we can say about it is that it does not run in with the general views of modern science as to the beginning of things, and that those who refuse to accept it must hold that, under certain conditions which prevail, life begins by a very gradual process, similar to that by which forms suggesting growth seem to originate even under conditions so unfavorable as those existing in a bottle of acid.

But it is not at all necessary for our purpose to decide this question. If life existed through a creative act, it is absurd to suppose that that act was confined to one of the countless millions of worlds scattered through space. If it began at a certain stage of evolution by a natural process, the question will arise, what conditions are favorable to the commencement of this process? Here we are quite justified in reasoning from what, granting this process, has taken place upon our globe during its past history. One of the most elementary principles accepted by the human mind is that like causes produce like effects. The special conditions under which we find life to develop around us may be comprehensively summed up as the existence of water in the liquid form, and the presence of nitrogen, free perhaps in the first place, but accompanied by substances with which it may form combinations. Oxygen, hydrogen, and nitrogen are, then, the fundamental requirements. The addition of calcium or other forms of matter necessary to the existence of a solid world goes without saying. The question now is whether these necessary conditions exist in other parts of the universe.

The spectroscope shows that, so far as the chemical elements go, other worlds are composed of the same elements as ours. Hydrogen especially exists everywhere, and we have reason to believe that the same is true of oxygen and nitrogen. Calcium, the base of lime, is almost universal. So far as chemical elements go, we may therefore take it for granted that the conditions under which life begins are very widely diffused in the universe. It is, therefore, contrary to all the analogies of nature to suppose that life began only on a single world.

It is a scientific inference, based on facts so numerous as not to admit of serious question, that during the history of our globe there has been a continually improving development of life. As ages upon ages pass, new forms are generated, higher in the scale than those which preceded them, until at length reason appears and asserts its sway. In a recent well-known work Alfred Russel Wallace has argued that this development of life required the presence of such a rare combination of conditions that there is no reason to suppose that it prevailed anywhere except on our earth. It is quite impossible in the present discussion to follow his reasoning in detail; but it seems to me altogether inconclusive. Not only does life, but intelligence, flourish on this globe under a great variety of conditions as regards temperature and surroundings, and no sound reason can be shown why under certain conditions, which are frequent in the universe, intelligent beings should not acquire the highest development.

Now let us look at the subject from the view of the mathematical theory of probabilities. A fundamental tenet of this theory is that no matter how improbable a result may be on a single trial, supposing it at all possible, it is sure to occur after a sufficient number of trials—and over and over again if the trials are repeated often enough. For example, if a million grains of corn, of which a single one was red, were all placed in a pile, and a blindfolded person were required to grope in the pile, select a grain, and then put it back again, the chances would be a million to one against his drawing out the red grain. If drawing it meant he should die, a sensible person would give himself no concern at having to draw the grain. The probability of his death would not be so great as the actual probability that he will really die within the next twenty-four hours. And yet if the whole human race were required to run this chance, it is certain that about fifteen hundred, or one out of a million, of the whole human family would draw the red grain and meet his death.

Now apply this principle to the universe. Let us suppose, to fix the ideas, that there are a hundred million worlds, but that the chances are one thousand to one against any one of these taken at random being fitted for the highest development of life or for the evolution of reason. The chances would still be that one hundred thousand of them would be inhabited by rational beings whom we call human. But where are we to look for these worlds? This no man can tell. We only infer from the statistics of the stars—and this inference is fairly well grounded—that the number of worlds which, so far as we know, may be inhabited, are to be counted by thousands, and perhaps by millions.

In a number of bodies so vast we should expect every variety of conditions as regards temperature and surroundings. If we suppose that the special conditions which prevail on our planet are necessary to the highest forms of life, we still have reason to believe that these same conditions prevail on thousands of other worlds. The fact that we might find the conditions in millions of other worlds unfavorable to life would not disprove the existence of the latter on countless worlds differently situated.

Coming down now from the general question to the specific one, we all know that the only worlds the conditions of which can be made the subject of observation are the planets which revolve around the sun, and their satellites. The question whether these bodies are inhabited is one which, of course, completely transcends not only our powers of observation at present, but every appliance of research that we can conceive of men devising. If Mars is inhabited, and if the people of that planet have equal powers with ourselves, the problem of merely producing an illumination which could be seen in our most powerful telescope would be beyond all the ordinary efforts of an entire nation. An unbroken square mile of flame would be invisible in our telescopes, but a hundred square miles might be seen. We cannot, therefore, expect to see any signs of the works of inhabitants even on Mars. All that we can do is to ascertain with greater or less probability whether the conditions necessary to life exist on the other planets of the system.

The moon being much the nearest to us of all the heavenly bodies, we can pronounce more definitely in its case than in any other. We know that neither air nor water exists on the moon in quantities sufficient to be perceived by the most delicate tests at our command. It is certain that the moon's atmosphere, if any exists, is less than the thousandth part of the density of that around us. The vacuum is greater than any ordinary air-pump is capable of producing. We can hardly suppose that so small a quantity of air could be of any benefit whatever in sustaining life; an animal that could get along on so little could get along on none at all.

But the proof of the absence of life is yet stronger when we consider the results of actual telescopic observation. An object such as an ordinary city block could be detected on the moon. If anything like vegetation were present on its surface, we should see the changes which it would undergo in the course of a month, during one portion of which it would be exposed to the rays of the unclouded sun, and during another to the intense cold of space. If men built cities, or even separate buildings the size of the larger ones on our earth, we might see some signs of them.

In recent times we not only observe the moon with the telescope, but get still more definite information by photography. The whole visible surface has been repeatedly photographed under the best conditions. But no change has been established beyond question, nor does the photograph show the slightest difference of structure or shade which could be attributed to cities or other works of man. To all appearances the whole surface of our satellite is as completely devoid of life as the lava newly thrown from Vesuvius. We next pass to the planets. Mercury, the nearest to the sun, is in a position very unfavorable for observation from the earth, because when nearest to us it is between us and the sun, so that its dark hemisphere is presented to us. Nothing satisfactory has yet been made out as to its condition. We cannot say with certainty whether it has an atmosphere or not. What seems very probable is that the temperature on its surface is higher than any of our earthly animals could sustain. But this proves nothing.

We know that Venus has an atmosphere. This was very conclusively shown during the transits of Venus in 1874 and 1882. But this atmosphere is so filled with clouds or vapor that it does not seem likely that we ever get a view of the solid body of the planet through it. Some observers have thought they could see spots on Venus day after day, while others have disputed this view. On the whole, if intelligent inhabitants live there, it is not likely that they ever see sun or stars. Instead of the sun they see only an effulgence in the vapory sky which disappears and reappears at regular intervals.

When we come to Mars, we have more definite knowledge, and there seems to be greater possibilities for life there than in the case of any other planet besides the earth. The main reason for denying that life such as ours could exist there is that the atmosphere of Mars is so rare that, in the light of the most recent researches, we cannot be fully assured that it exists at all. The very careful comparisons of the spectra of Mars and of the moon made by Campbell at the Lick Observatory failed to show the slightest difference in the two. If Mars had an atmosphere as dense as ours, the result could be seen in the darkening of the lines of the spectrum produced by the double passage of the light through it. There were no lines in the spectrum of Mars that were not seen with equal distinctness in that of the moon. But this does not prove the entire absence of an atmosphere. It only shows a limit to its density. It may be one-fifth or one-fourth the density of that on the earth, but probably no more.

That there must be something in the nature of vapor at least seems to be shown by the formation and disappearance of the white polar caps of this planet. Every reader of astronomy at the present time knows that, during the Martian winter, white caps form around the pole of the planet which is turned away from the sun, and grow larger and larger until the sun begins to shine upon them, when they gradually grow smaller, and perhaps nearly disappear. It seems, therefore, fairly well proved that, under the influence of cold, some white substance forms around the polar regions of Mars which evaporates under the influence of the sun's rays. It has been supposed that this substance is snow, produced in the same way that snow is produced on the earth, by the evaporation of water.

But there are difficulties in the way of this explanation. The sun sends less than half as much heat to Mars as to the earth, and it does not seem likely that the polar regions can ever receive enough of heat to melt any considerable quantity of snow. Nor does it seem likely that any clouds from which snow could fall ever obscure the surface of Mars.

But a very slight change in the explanation will make it tenable. Quite possibly the white deposits may be due to something like hoar-frost condensed from slightly moist air, without the actual production of snow. This would produce the effect that we see. Even this explanation implies that Mars has air and water, rare though the former may be. It is quite possible that air as thin as that of Mars would sustain life in some form. Life not totally unlike that on the earth may therefore exist upon this planet for anything that we know to the contrary. More than this we cannot say.

In the case of the outer planets the answer to our question must be in the negative. It now seems likely that Jupiter is a body very much like our sun, only that the dark portion is too cool to emit much, if any, light. It is doubtful whether Jupiter has anything in the nature of a solid surface. Its interior is in all likelihood a mass of molten matter far above a red heat, which is surrounded by a comparatively cool, yet, to our measure, extremely hot, vapor. The belt-like clouds which surround the planet are due to this vapor combined with the rapid rotation. If there is any solid surface below the atmosphere that we can see, it is swept by winds such that nothing we have on earth could withstand them. But, as we have said, the probabilities are very much against there being anything like such a surface. At some great depth in the fiery vapor there is a solid nucleus; that is all we can say.

The planet Saturn seems to be very much like that of Jupiter in its composition. It receives so little heat from the sun that, unless it is a mass of fiery vapor like Jupiter, the surface must be far below the freezing-point.

We cannot speak with such certainty of Uranus and Neptune; yet the probability seems to be that they are in much the same condition as Saturn. They are known to have very dense atmospheres, which are made known to us only by their absorbing some of the light of the sun. But nothing is known of the composition of these atmospheres.

To sum up our argument: the fact that, so far as we have yet been able to learn, only a very small proportion of the visible worlds scattered through space are fitted to be the abode of life does not preclude the probability that among hundreds of millions of such worlds a vast number are so fitted.

Such being the case, all the analogies of nature lead us to believe that, whatever the process which led to life upon this earth—whether a special act of creative power or a gradual course of development—through that same process does life begin in every part of the universe fitted to sustain it. The course of development involves a gradual improvement in living forms, which by irregular steps rise higher and higher in the scale of being. We have every reason to believe that this is the case wherever life exists. It is, therefore, perfectly reasonable to suppose that beings, not only animated, but endowed with reason, inhabit countless worlds in space. It would, indeed, be very inspiring could we learn by actual observation what forms of society exist throughout space, and see the members of such societies enjoying themselves by their warm firesides. But this, so far as we can now see, is entirely beyond the possible reach of our race, so long as it is confined to a single world.

HOW THE PLANETS ARE WEIGHED

You ask me how the planets are weighed? I reply, on the same principle by which a butcher weighs a ham in a spring-balance. When he picks the ham up, he feels a pull of the ham towards the earth. When he hangs it on the hook, this pull is transferred from his hand to the spring of the balance. The stronger the pull, the farther the spring is pulled down. What he reads on the scale is the strength of the pull. You know that this pull is simply the attraction of the earth on the ham. But, by a universal law of force, the ham attracts the earth exactly as much as the earth does the ham. So what the butcher really does is to find how much or how strongly the ham attracts the earth, and he calls that pull the weight of the ham. On the same principle, the astronomer finds the weight of a body by finding how strong is its attractive pull on some other body. If the butcher, with his spring-balance and a ham, could fly to all the planets, one after the other, weigh the ham on each, and come back to report the results to an astronomer, the latter could immediately compute the weight of each planet of known diameter, as compared with that of the earth. In applying this principle to the heavenly bodies, we at once meet a difficulty that looks insurmountable. You cannot get up to the heavenly bodies to do your weighing; how then will you measure their pull? I must begin the answer to this question by explaining a nice point in exact science. Astronomers distinguish between the weight of a body and its mass. The weight of objects is not the same all over the world; a thing which weighs thirty pounds in New York would weigh an ounce more than thirty pounds in a spring-balance in Greenland, and nearly an ounce less at the equator. This is because the earth is not a perfect sphere, but a little flattened. Thus weight varies with the place. If a ham weighing thirty pounds were taken up to the moon and weighed there, the pull would only be five pounds, because the moon is so much smaller and lighter than the earth. There would be another weight of the ham for the planet Mars, and yet another on the sun, where it would weigh some eight hundred pounds. Hence the astronomer does not speak of the weight of a planet, because that would depend on the place where it was weighed; but he speaks of the mass of the planet, which means how much planet there is, no matter where you might weigh it.

At the same time, we might, without any inexactness, agree that the mass of a heavenly body should be fixed by the weight it would have in New York. As we could not even imagine a planet at New York, because it may be larger than the earth itself, what we are to imagine is this: Suppose the planet could be divided into a million million million equal parts, and one of these parts brought to New York and weighed. We could easily find its weight in pounds or tons. Then multiply this weight by a million million million, and we shall have a weight of the planet. This would be what the astronomers might take as the mass of the planet.

With these explanations, let us see how the weight of the earth is found. The principle we apply is that round bodies of the same specific gravity attract small objects on their surface with a force proportional to the diameter of the attracting body. For example, a body two feet in diameter attracts twice as strongly as one of a foot, one of three feet three times as strongly, and so on. Now, our earth is about 40,000,000 feet in diameter; that is 10,000,000 times four feet. It follows that if we made a little model of the earth four feet in diameter, having the average specific gravity of the earth, it would attract a particle with one ten-millionth part of the attraction of the earth. The attraction of such a model has actually been measured. Since we do not know the average specific gravity of the earth—that being in fact what we want to find out—we take a globe of lead, four feet in diameter, let us suppose. By means of a balance of the most exquisite construction it is found that such a globe does exert a minute attraction on small bodies around it, and that this attraction is a little more than the ten-millionth part of that of the earth. This shows that the specific gravity of the lead is a little greater than that of the average of the whole earth. All the minute calculations made, it is found that the earth, in order to attract with the force it does, must be about five and one-half times as heavy as its bulk of water, or perhaps a little more. Different experimenters find different results; the best between 5.5 and 5.6, so that 5.5 is, perhaps, as near the number as we can now get. This is much more than the average specific gravity of the materials which compose that part of the earth which we can reach by digging mines. The difference arises from the fact that, at the depth of many miles, the matter composing the earth is compressed into a smaller space by the enormous weight of the portions lying above it. Thus, at the depth of 1000 miles, the pressure on every cubic inch is more than 2000 tons, a weight which would greatly condense the hardest metal.

We come now to the planets. I have said that the mass or weight of a heavenly body is determined by its attraction on some other body. There are two ways in which the attraction of a planet may be measured. One is by its attraction on the planets next to it. If these bodies did not attract one another at all, but only moved under the influence of the sun, they would move in orbits having the form of ellipses. They are found to move very nearly in such orbits, only the actual path deviates from an ellipse, now in one direction and then in another, and it slowly changes its position from year to year. These deviations are due to the pull of the other planets, and by measuring the deviations we can determine the amount of the pull, and hence the mass of the planet.

The reader will readily understand that the mathematical processes necessary to get a result in this way must be very delicate and complicated. A much simpler method can be used in the case of those planets which have satellites revolving round them, because the attraction of the planet can be determined by the motions of the satellite. The first law of motion teaches us that a body in motion, if acted on by no force, will move in a straight line. Hence, if we see a body moving in a curve, we know that it is acted on by a force in the direction towards which the motion curves. A familiar example is that of a stone thrown from the hand. If the stone were not attracted by the earth, it would go on forever in the line of throw, and leave the earth entirely. But under the attraction of the earth, it is drawn down and down, as it travels onward, until finally it reaches the ground. The faster the stone is thrown, of course, the farther it will go, and the greater will be the sweep of the curve of its path. If it were a cannon-ball, the first part of the curve would be nearly a right line. If we could fire a cannon-ball horizontally from the top of a high mountain with a velocity of five miles a second, and if it were not resisted by the air, the curvature of the path would be equal to that of the surface of our earth, and so the ball would never reach the earth, but would revolve round it like a little satellite in an orbit of its own. Could this be done, the astronomer would be able, knowing the velocity of the ball, to calculate the attraction of the earth as well as we determine it by actually observing the motion of falling bodies around us.

Thus it is that when a planet, like Mars or Jupiter, has satellites revolving round it, astronomers on the earth can observe the attraction of the planet on its satellites and thus determine its mass. The rule for doing this is very simple. The cube of the distance between the planet and satellite is divided by the square of the time of revolution of the satellite. The quotient is a number which is proportional to the mass of the planet. The rule applies to the motion of the moon round the earth and of the planets round the sun. If we divide the cube of the earth's distance from the sun, say 93,000,000 miles, by the square of 365 1/4, the days in a year, we shall get a certain quotient. Let us call this number the sun-quotient. Then, if we divide the cube of the moon's distance from the earth by the square of its time of revolution, we shall get another quotient, which we may call the earth-quotient. The sun-quotient will come out about 330,000 times as large as the earth-quotient. Hence it is concluded that the mass of the sun is 330,000 times that of the earth; that it would take this number of earths to make a body as heavy as the sun.

I give this calculation to illustrate the principle; it must not be supposed that the astronomer proceeds exactly in this way and has only this simple calculation to make. In the case of the moon and earth, the motion and distance of the former vary in consequence of the attraction of the sun, so that their actual distance apart is a changing quantity. So what the astronomer actually does is to find the attraction of the earth by observing the length of a pendulum which beats seconds in various latitudes. Then, by very delicate mathematical processes, he can find with great exactness what would be the time of revolution of a small satellite at any given distance from the earth, and thus can get the earth-quotient.

But, as I have already pointed out, we must, in the case of the planets, find the quotient in question by means of the satellites; and it happens, fortunately, that the motions of these bodies are much less changed by the attraction of the sun than is the motion of the moon. Thus, when we make the computation for the outer satellite of Mars, we find the quotient to be 1/3093500 that of the sun-quotient. Hence we conclude that the mass of Mars is 1/3093500 that of the sun. By the corresponding quotient, the mass of Jupiter is found to be about 1/1047 that of the sun, Saturn 1/3500, Uranus 1/22700, Neptune 1/19500.

We have set forth only the great principle on which the astronomer has proceeded for the purpose in question. The law of gravitation is at the bottom of all his work. The effects of this law require mathematical processes which it has taken two hundred years to bring to their present state, and which are still far from perfect. The measurement of the distance of a satellite is not a job to be done in an evening; it requires patient labor extending through months and years, and then is not as exact as the astronomer would wish. He does the best he can, and must be satisfied with that.

THE MARINER'S COMPASS

Among those provisions of Nature which seem to us as especially designed for the use of man, none is more striking than the seeming magnetism of the earth. What would our civilization have been if the mariner's compass had never been known? That Columbus could never have crossed the Atlantic is certain; in what generation since his time our continent would have been discovered is doubtful. Did the reader ever reflect what a problem the captain of the finest ocean liner of our day would face if he had to cross the ocean without this little instrument? With the aid of a pilot he gets his ship outside of Sandy Hook without much difficulty. Even later, so long as the sun is visible and the air is clear, he will have some apparatus for sailing by the direction of the sun. But after a few hours clouds cover the sky. From that moment he has not the slightest idea of east, west, north, or south, except so far as he may infer it from the direction in which he notices the wind to blow. For a few hours he may be guided by the wind, provided he is sure he is not going ashore on Long Island. Thus, in time, he feels his way out into the open sea. By day he has some idea of direction with the aid of the sun; by night, when the sky is clear he can steer by the Great Bear, or "Cynosure," the compass of his ancient predecessors on the Mediterranean. But when it is cloudy, if he persists in steaming ahead, he may be running towards the Azores or towards Greenland, or he may be making his way back to New York without knowing it. So, keeping up steam only when sun or star is visible, he at length finds that he is approaching the coast of Ireland. Then he has to grope along much like a blind man with his staff, feeling his way along the edge of a precipice. He can determine the latitude at noon if the sky is clear, and his longitude in the morning or evening in the same conditions. In this way he will get a general idea of his whereabouts. But if he ventures to make headway in a fog, he may find himself on the rocks at any moment. He reaches his haven only after many spells of patient waiting for favoring skies.

The fact that the earth acts like a magnet, that the needle points to the north, has been generally known to navigators for nearly a thousand years, and is said to have been known to the Chinese at a yet earlier period. And yet, to-day, if any professor of physical science is asked to explain the magnetic property of the earth, he will acknowledge his inability to do so to his own satisfaction. Happily this does not hinder us from finding out by what law these forces act, and how they enable us to navigate the ocean. I therefore hope the reader will be interested in a short exposition of the very curious and interesting laws on which the science of magnetism is based, and which are applied in the use of the compass.

The force known as magnetic, on which the compass depends, is different from all other natural forces with which we are familiar. It is very remarkable that iron is the only substance which can become magnetic in any considerable degree. Nickel and one or two other metals have the same property, but in a very slight degree. It is also remarkable that, however powerfully a bar of steel may be magnetized, not the slightest effect of the magnetism can be seen by its action on other than magnetic substances. It is no heavier than before. Its magnetism does not produce the slightest influence upon the human body. No one would know that it was magnetic until something containing iron was brought into its immediate neighborhood; then the attraction is set up. The most important principle of magnetic science is that there are two opposite kinds of magnetism, which are, in a certain sense, contrary in their manifestations. The difference is seen in the behavior of the magnet itself. One particular end points north, and the other end south. What is it that distinguishes these two ends? The answer is that one end has what we call north magnetism, while the other has south magnetism. Every magnetic bar has two poles, one near one end, one near the other. The north pole is drawn towards the north pole of the earth, the south pole towards the south pole, and thus it is that the direction of the magnet is determined. Now, when we bring two magnets near each other we find another curious phenomenon. If the two like poles are brought together, they do not attract but repel each other. But the two opposite poles attract each other. The attraction and repulsion are exactly equal under the same conditions. There is no more attraction than repulsion. If we seal one magnet up in a paper or a box, and then suspend another over the box, the north pole of the one outside will tend to the south pole of the one in the box, and vice versa.

Our next discovery is, that whenever a magnet attracts a piece of iron it makes that iron into a magnet, at least for the time being. In the case of ordinary soft or untempered iron the magnetism disappears instantly when the magnet is removed. But if the magnet be made to attract a piece of hardened steel, the latter will retain the magnetism produced in it and become itself a permanent magnet.

This fact must have been known from the time that the compass came into use. To make this instrument it was necessary to magnetize a small bar or needle by passing a natural magnet over it.

In our times the magnetization is effected by an electric current. The latter has curious magnetic properties; a magnetic needle brought alongside of it will be found placing itself at right angles to the wire bearing the current. On this principle is made the galvanometer for measuring the intensity of a current. Moreover, if a piece of wire is coiled round a bar of steel, and a powerful electric current pass through the coil, the bar will become a magnet.

Another curious property of magnetism is that we cannot develop north magnetism in a bar without developing south magnetism at the same time. If it were otherwise, important consequences would result. A separate north pole of a magnet would, if attached to a floating object and thrown into the ocean, start on a journey towards the north all by itself. A possible method of bringing this result about may suggest itself. Let us take an ordinary bar magnet, with a pole at each end, and break it in the middle; then would not the north end be all ready to start on its voyage north, and the south end to make its way south? But, alas! when this experiment is tried it is found that a south pole instantly develops itself on one side of the break, and a north pole on the other side, so that the two pieces will simply form two magnets, each with its north and south pole. There is no possibility of making a magnet with only one pole.

It was formerly supposed that the central portions of the earth consisted of an immense magnet directed north and south. Although this view is found, for reasons which need not be set forth in detail, to be untenable, it gives us a good general idea of the nature of terrestrial magnetism. One result that follows from the law of poles already mentioned is that the magnetism which seems to belong to the north pole of the earth is what we call south on the magnet, and vice versa.

Careful experiment shows us that the region around every magnet is filled with magnetic force, strongest near the poles of the magnet, but diminishing as the inverse square of the distance from the pole. This force, at each point, acts along a certain line, called a line of force. These lines are very prettily shown by the familiar experiment of placing a sheet of paper over a magnet, and then scattering iron filings on the surface of the paper. It will be noticed that the filings arrange themselves along a series of curved lines, diverging in every direction from each pole, but always passing from one pole to the other. It is a universal law that whenever a magnet is brought into a region where this force acts, it is attracted into such a position that it shall have the same direction as the lines of force. Its north pole will take the direction of the curve leading to the south pole of the other magnet, and its south pole the opposite one.

The fact of terrestrial magnetism may be expressed by saying that the space within and around the whole earth is filled by lines of magnetic force, which we know nothing about until we suspend a magnet so perfectly balanced that it may point in any direction whatever. Then it turns and points in the direction of the lines of force, which may thus be mapped out for all points of the earth.

We commonly say that the pole of the needle points towards the north. The poets tell us how the needle is true to the pole. Every reader, however, is now familiar with the general fact of a variation of the compass. On our eastern seaboard, and all the way across the Atlantic, the north pointing of the compass varies so far to the west that a ship going to Europe and making no allowance for this deviation would find herself making more nearly for the North Cape than for her destination. The "declination," as it is termed in scientific language, varies from one region of the earth to another. In some places it is towards the west, in others towards the east.

The pointing of the needle in various regions of the world is shown by means of magnetic maps. Such maps are published by the United States Coast Survey, whose experts make a careful study of the magnetic force all over the country. It is found that there is a line running nearly north and south through the Middle States along which there is no variation of the compass. To the east of it the variation of the north pole of the magnet is west; to the west of it, east. The most rapid changes in the pointing of the needle are towards the northeast and northwest regions. When we travel to the northeastern boundary of Maine the westerly variation has risen to 20 degrees. Towards the northwest the easterly variation continually increases, until, in the northern part of the State of Washington, it amounts to 23 degrees.

When we cross the Atlantic into Europe we find the west variation diminishing until we reach a certain line passing through central Russia and western Asia. This is again a line of no variation. Crossing it, the variation is once more towards the east. This direction continues over most of the continent of Asia, but varies in a somewhat irregular manner from one part of the continent to another.

As a general rule, the lines of the earth's magnetic force are not horizontal, and therefore one end or the other of a perfectly suspended magnet will dip below the horizontal position. This is called the "dip of the needle." It is observed by means of a brass circle, of which the circumference is marked off in degrees. A magnet is attached to this circle so as to form a diameter, and suspended on a horizontal axis passing through the centre of gravity, so that the magnet shall be free to point in the direction indicated by the earth's lines of magnetic force. Armed with this apparatus, scientific travellers and navigators have visited various points of the earth in order to determine the dip. It is thus found that there is a belt passing around the earth near the equator, but sometimes deviating several degrees from it, in which there is no dip; that is to say, the lines of magnetic force are horizontal. Taking any point on this belt and going north, it will be found that the north pole of the magnet gradually tends downward, the dip constantly increasing as we go farther north. In the southern part of the United States the dip is about 60 degrees, and the direction of the needle is nearly perpendicular to the earth's axis. In the northern part of the country, including the region of the Great Lakes, the dip increases to 75 degrees. Noticing that a dip of 90 degrees would mean that the north end of the magnet points straight downward, it follows that it would be more nearly correct to say that, throughout the United States, the magnetic needle points up and down than that it points north and south.

Going yet farther north, we find the dip still increasing, until at a certain point in the arctic regions the north pole of the needle points downward. In this region the compass is of no use to the traveller or the navigator. The point is called the Magnetic Pole. Its position has been located several times by scientific observers. The best determinations made during the last eighty years agree fairly well in placing it near 70 degrees north latitude and 97 degrees longitude west from Greenwich. This point is situated on the west shore of the Boothian Peninsula, which is bounded on the south end by McClintock Channel. It is about five hundred miles north of the northwest part of Hudson Bay. There is a corresponding magnetic pole in the Antarctic Ocean, or rather on Victoria Land, nearly south of Australia. Its position has not been so exactly located as in the north, but it is supposed to be at about 74 degrees of south latitude and 147 degrees of east longitude from Greenwich.

The magnetic poles used to be looked upon as the points towards which the respective ends of the needle were attracted. And, as a matter of fact, the magnetic force is stronger near the poles than elsewhere. When located in this way by strength of force, it is found that there is a second north pole in northern Siberia. Its location has not, however, been so well determined as in the case of the American pole, and it is not yet satisfactorily shown that there is any one point in Siberia where the direction of the force is exactly downward.

The declination and dip, taken together, show the exact direction of the magnetic force at any place. But in order to complete the statement of the force, one more element must be given—its amount. The intensity of the magnetic force is determined by suspending a magnet in a horizontal position, and then allowing it to oscillate back and forth around the suspension. The stronger the force, the less the time it will take to oscillate. Thus, by carrying a magnet to various parts of the world, the magnetic force can be determined at every point where a proper support for the magnet is obtainable. The intensity thus found is called the horizontal force. This is not really the total force, because the latter depends upon the dip; the greater the dip, the less will be the horizontal force which corresponds to a certain total force. But a very simple computation enables the one to be determined when the value of the other is known. In this way it is found that, as a general rule, the magnetic force is least in the earth's equatorial regions and increases as we approach either of the magnetic poles.

When the most exact observations on the direction of the needle are made, it is found that it never remains at rest. Beginning with the changes of shortest duration, we have a change which takes place every day, and is therefore called diurnal. In our northern latitudes it is found that during the six hours from nine o'clock at night until three in the morning the direction of the magnet remains nearly the same. But between three and four A.M. it begins to deviate towards the east, going farther and farther east until about 8 A.M. Then, rather suddenly, it begins to swing towards the west with a much more rapid movement, which comes to an end between one and two o'clock in the afternoon. Then, more slowly, it returns in an easterly direction until about nine at night, when it becomes once more nearly quiescent. Happily, the amount of this change is so small that the navigator need not trouble himself with it. The entire range of movement rarely amounts to one-quarter of a degree.

It is a curious fact that the amount of the change is twice as great in June as it is in December. This indicates that it is caused by the sun's radiation. But how or why this cause should produce such an effect no one has yet discovered.

Another curious feature is that in the southern hemisphere the direction of the motion is reversed, although its general character remains the same. The pointing deviates towards the west in the morning, then rapidly moves towards the east until about two o'clock, after which it slowly returns to its original direction.

The dip of the needle goes through a similar cycle of daily changes. In northern latitudes it is found that at about six in the morning the dip begins to increase, and continues to do so until noon, after which it diminishes until seven or eight o'clock in the evening, when it becomes nearly constant for the rest of the night. In the southern hemisphere the direction of the movement is reversed.

When the pointing of the needle is compared with the direction of the moon, it is found that there is a similar change. But, instead of following the moon in its course, it goes through two periods in a day, like the tides. When the moon is on the meridian, whether above or below us, the effect is in one direction, while when it is rising or setting it is in the opposite direction. In other words, there is a complete swinging backward and forward twice in a lunar day. It might be supposed that such an effect would be due to the moon, like the earth, being a magnet. But were this the case there would be only one swing back and forth during the passage of the moon from the meridian until it came back to the meridian again. The effect would be opposite at the rising and setting of the moon, which we have seen is not the case. To make the explanation yet more difficult, it is found that, as in the case of the sun, the change is opposite in the northern and southern hemispheres and very small at the equator, where, by virtue of any action that we can conceive of, it ought to be greatest. The pointing is also found to change with the age of the moon and with the season of the year. But these motions are too small to be set forth in the present article.

There is yet another class of changes much wider than these. The observations recorded since the time of Columbus show that, in the course of centuries, the variation of the compass, at any one point, changes very widely. It is well known that in 1490 the needle pointed east of north in the Mediterranean, as well as in those portions of the Atlantic which were then navigated. Columbus was therefore much astonished when, on his first voyage, in mid-ocean, he found that the deviation was reversed, and was now towards the west. It follows that a line of no variation then passed through the Atlantic Ocean. But this line has since been moving towards the east. About 1662 it passed the meridian of Paris. During the two hundred and forty years which have since elapsed, it has passed over Central Europe, and now, as we have already said, passes through European Russia.

The existence of natural magnets composed of iron ore, and their property of attracting iron and making it magnetic, have been known from the remotest antiquity. But the question as to who first discovered the fact that a magnetized needle points north and south, and applied this discovery to navigation, has given rise to much discussion. That the property was known to the Chinese about the beginning of our era seems to be fairly well established, the statements to that effect being of a kind that could not well have been invented. Historical evidence of the use of the magnetic needle in navigation dates from the twelfth century. The earliest compass consisted simply of a splinter of wood or a piece of straw to which the magnetized needle was attached, and which was floated in water. A curious obstacle is said to have interfered with the first uses of this instrument. Jack is a superstitious fellow, and we may be sure that he was not less so in former times than he is today. From his point of view there was something uncanny in so very simple a contrivance as a floating straw persistently showing him the direction in which he must sail. It made him very uncomfortable to go to sea under the guidance of an invisible power. But with him, as with the rest of us, familiarity breeds contempt, and it did not take more than a generation to show that much good and no harm came to those who used the magic pointer.

The modern compass, as made in the most approved form for naval and other large ships, is the liquid one. This does not mean that the card bearing the needle floats on the liquid, but only that a part of the force is taken off from the pivot on which it turns, so as to make the friction as small as possible, and to prevent the oscillation back and forth which would continually go on if the card were perfectly free to turn. The compass-card is marked not only with the thirty-two familiar points of the compass, but is also divided into degrees. In the most accurate navigation it is probable that very little use of the points is made, the ship being directed according to the degrees.

A single needle is not relied upon to secure the direction of the card, the latter being attached to a system of four or even more magnets, all pointing in the same direction. The compass must have no iron in its construction or support, because the attraction of that substance on the needle would be fatal to its performance.

From this cause the use of iron as ship-building material introduced a difficulty which it was feared would prove very serious. The thousands of tons of iron in a ship must exert a strong attraction on the magnetic needle. Another complication is introduced by the fact that the iron of the ship will always become more or less magnetic, and when the ship is built of steel, as modern ones are, this magnetism will be more or less permanent.

We have already said that a magnet has the property of making steel or iron in its neighborhood into another magnet, with its poles pointing in the opposite direction. The consequence is that the magnetism of the earth itself will make iron or steel more or less magnetic. As a ship is built she thus becomes a great repository of magnetism, the direction of the force of which will depend upon the position in which she lay while building. If erected on the bank of an east and west stream, the north end of the ship will become the north pole of a magnet and the south end the south pole. Accordingly, when she is launched and proceeds to sea, the compass points not exactly according to the magnetism of the earth, but partly according to that of the ship also.

The methods of obviating this difficulty have exercised the ingenuity of the ablest physicists from the beginning of iron ship building. One method is to place in the neighborhood of the compass, but not too near it, a steel bar magnetized in the opposite direction from that of the ship, so that the action of the latter shall be neutralized. But a perfect neutralization cannot be thus effected. It is all the more difficult to effect it because the magnetism of a ship is liable to change.

The practical method therefore adopted is called "swinging the ship," an operation which passengers on ocean liners may have frequently noticed when approaching land. The ship is swung around so that her bow shall point in various directions. At each pointing the direction of the ship is noticed by sighting on the sun, and also the direction of the compass itself. In this way the error of the pointing of the compass as the ship swings around is found for every direction in which she may be sailing. A table can then be made showing what the pointing, according to the compass, should be in order that the ship may sail in any given direction.

This, however, does not wholly avoid the danger. The tables thus made are good when the ship is on a level keel. If, from any cause whatever, she heels over to one side, the action will be different. Thus there is a "heeling error" which must be allowed for. It is supposed to have been from this source of error not having been sufficiently determined or appreciated that the lamentable wreck of the United States ship Huron off the coast of Hatteras occurred some twenty years ago.

THE FAIRYLAND OF GEOMETRY

If the reader were asked in what branch of science the imagination is confined within the strictest limits, he would, I fancy, reply that it must be that of mathematics. The pursuer of this science deals only with problems requiring the most exact statements and the most rigorous reasoning. In all other fields of thought more or less room for play may be allowed to the imagination, but here it is fettered by iron rules, expressed in the most rigid logical form, from which no deviation can be allowed. We are told by philosophers that absolute certainty is unattainable in all ordinary human affairs, the only field in which it is reached being that of geometric demonstration.

And yet geometry itself has its fairyland—a land in which the imagination, while adhering to the forms of the strictest demonstration, roams farther than it ever did in the dreams of Grimm or Andersen. One thing which gives this field its strictly mathematical character is that it was discovered and explored in the search after something to supply an actual want of mathematical science, and was incited by this want rather than by any desire to give play to fancy. Geometricians have always sought to found their science on the most logical basis possible, and thus have carefully and critically inquired into its foundations. The new geometry which has thus arisen is of two closely related yet distinct forms. One of these is called NON-EUCLIDIAN, because Euclid's axiom of parallels, which we shall presently explain, is ignored. In the other form space is assumed to have one or more dimensions in addition to the three to which the space we actually inhabit is confined. As we go beyond the limits set by Euclid in adding a fourth dimension to space, this last branch as well as the other is often designated non-Euclidian. But the more common term is hypergeometry, which, though belonging more especially to space of more than three dimensions, is also sometimes applied to any geometric system which transcends our ordinary ideas.

In all geometric reasoning some propositions are necessarily taken for granted. These are called axioms, and are commonly regarded as self-evident. Yet their vital principle is not so much that of being self-evident as being, from the nature of the case, incapable of demonstration. Our edifice must have some support to rest upon, and we take these axioms as its foundation. One example of such a geometric axiom is that only one straight line can be drawn between two fixed points; in other words, two straight lines can never intersect in more than a single point. The axiom with which we are at present concerned is commonly known as the 11th of Euclid, and may be set forth in the following way: We have given a straight line, A B, and a point, P, with another line, C D, passing through it and capable of being turned around on P. Euclid assumes that this line C D will have one position in which it will be parallel to A B, that is, a position such that if the two lines are produced without end, they will never meet. His axiom is that only one such line can be drawn through P. That is to say, if we make the slightest possible change in the direction of the line C D, it will intersect the other line, either in one direction or the other.

The new geometry grew out of the feeling that this proposition ought to be proved rather than taken as an axiom; in fact, that it could in some way be derived from the other axioms. Many demonstrations of it were attempted, but it was always found, on critical examination, that the proposition itself, or its equivalent, had slyly worked itself in as part of the base of the reasoning, so that the very thing to be proved was really taken for granted.

This suggested another course of inquiry. If this axiom of parallels does not follow from the other axioms, then from these latter we may construct a system of geometry in which the axiom of parallels shall not be true. This was done by Lobatchewsky and Bolyai, the one a Russian the other a Hungarian geometer, about 1830.

To show how a result which looks absurd, and is really inconceivable by us, can be treated as possible in geometry, we must have recourse to analogy. Suppose a world consisting of a boundless flat plane to be inhabited by reasoning beings who can move about at pleasure on the plane, but are not able to turn their heads up or down, or even to see or think of such terms as above them and below them, and things around them can be pushed or pulled about in any direction, but cannot be lifted up. People and things can pass around each other, but cannot step over anything. These dwellers in "flatland" could construct a plane geometry which would be exactly like ours in being based on the axioms of Euclid. Two parallel straight lines would never meet, though continued indefinitely.

But suppose that the surface on which these beings live, instead of being an infinitely extended plane, is really the surface of an immense globe, like the earth on which we live. It needs no knowledge of geometry, but only an examination of any globular object—an apple, for example—to show that if we draw a line as straight as possible on a sphere, and parallel to it draw a small piece of a second line, and continue this in as straight a line as we can, the two lines will meet when we proceed in either direction one-quarter of the way around the sphere. For our "flat-land" people these lines would both be perfectly straight, because the only curvature would be in the direction downward, which they could never either perceive or discover. The lines would also correspond to the definition of straight lines, because any portion of either contained between two of its points would be the shortest distance between those points. And yet, if these people should extend their measures far enough, they would find any two parallel lines to meet in two points in opposite directions. For all small spaces the axioms of their geometry would apparently hold good, but when they came to spaces as immense as the semi-diameter of the earth, they would find the seemingly absurd result that two parallel lines would, in the course of thousands of miles, come together. Another result yet more astonishing would be that, going ahead far enough in a straight line, they would find that although they had been going forward all the time in what seemed to them the same direction, they would at the end of 25,000 miles find themselves once more at their starting-point.

One form of the modern non-Euclidian geometry assumes that a similar theorem is true for the space in which our universe is contained. Although two straight lines, when continued indefinitely, do not appear to converge even at the immense distances which separate us from the fixed stars, it is possible that there may be a point at which they would eventually meet without either line having deviated from its primitive direction as we understand the case. It would follow that, if we could start out from the earth and fly through space in a perfectly straight line with a velocity perhaps millions of times that of light, we might at length find ourselves approaching the earth from a direction the opposite of that in which we started. Our straight-line circle would be complete.

Another result of the theory is that, if it be true, space, though still unbounded, is not infinite, just as the surface of a sphere, though without any edge or boundary, has only a limited extent of surface. Space would then have only a certain volume—a volume which, though perhaps greater than that of all the atoms in the material universe, would still be capable of being expressed in cubic miles. If we imagine our earth to grow larger and larger in every direction without limit, and with a speed similar to that we have described, so that to-morrow it was large enough to extend to the nearest fixed stars, the day after to yet farther stars, and so on, and we, living upon it, looked out for the result, we should, in time, see the other side of the earth above us, coming down upon us? as it were. The space intervening would grow smaller, at last being filled up. The earth would then be so expanded as to fill all existing space.

This, although to us the most interesting form of the non-Euclidian geometry, is not the only one. The idea which Lobatchewsky worked out was that through a point more than one parallel to a given line could be drawn; that is to say, if through the point P we have already supposed another line were drawn making ever so small an angle with CD, this line also would never meet the line AB. It might approach the latter at first, but would eventually diverge. The two lines AB and CD, starting parallel, would eventually, perhaps at distances greater than that of the fixed stars, gradually diverge from each other. This system does not admit of being shown by analogy so easily as the other, but an idea of it may be had by supposing that the surface of "flat-land," instead of being spherical, is saddle-shaped. Apparently straight parallel lines drawn upon it would then diverge, as supposed by Bolyai. We cannot, however, imagine such a surface extended indefinitely without losing its properties. The analogy is not so clearly marked as in the other case.

To explain hypergeometry proper we must first set forth what a fourth dimension of space means, and show how natural the way is by which it may be approached. We continue our analogy from "flat-land" In this supposed land let us make a cross—two straight lines intersecting at right angles. The inhabitants of this land understand the cross perfectly, and conceive of it just as we do. But let us ask them to draw a third line, intersecting in the same point, and perpendicular to both the other lines. They would at once pronounce this absurd and impossible. It is equally absurd and impossible to us if we require the third line to be drawn on the paper. But we should reply, "If you allow us to leave the paper or flat surface, then we can solve the problem by simply drawing the third line through the paper perpendicular to its surface."

Now, to pursue the analogy, suppose that, after we have drawn three mutually perpendicular lines, some being from another sphere proposes to us the drawing of a fourth line through the same point, perpendicular to all three of the lines already there. We should answer him in the same way that the inhabitants of "flat-land" answered us: "The problem is impossible. You cannot draw any such line in space as we understand it." If our visitor conceived of the fourth dimension, he would reply to us as we replied to the "flat-land" people: "The problem is absurd and impossible if you confine your line to space as you understand it. But for me there is a fourth dimension in space. Draw your line through that dimension, and the problem will be solved. This is perfectly simple to me; it is impossible to you solely because your conceptions do not admit of more than three dimensions."

Supposing the inhabitants of "flat-land" to be intellectual beings as we are, it would be interesting to them to be told what dwellers of space in three dimensions could do. Let us pursue the analogy by showing what dwellers in four dimensions might do. Place a dweller of "flat-land" inside a circle drawn on his plane, and ask him to step outside of it without breaking through it. He would go all around, and, finding every inch of it closed, he would say it was impossible from the very nature of the conditions. "But," we would reply, "that is because of your limited conceptions. We can step over it."

"Step over it!" he would exclaim. "I do not know what that means. I can pass around anything if there is a way open, but I cannot imagine what you mean by stepping over it."

But we should simply step over the line and reappear on the other side. So, if we confine a being able to move in a fourth dimension in the walls of a dungeon of which the sides, the floor, and the ceiling were all impenetrable, he would step outside of it without touching any part of the building, just as easily as we could step over a circle drawn on the plane without touching it. He would simply disappear from our view like a spirit, and perhaps reappear the next moment outside the prison. To do this he would only have to make a little excursion in the fourth dimension.

Another curious application of the principle is more purely geometrical. We have here two triangles, of which the sides and angles of the one are all equal to corresponding sides and angles of the other. Euclid takes it for granted that the one triangle can be laid upon the other so that the two shall fit together. But this cannot be done unless we lift one up and turn it over. In the geometry of "flat-land" such a thing as lifting up is inconceivable; the two triangles could never be fitted together.

Now let us suppose two pyramids similarly related. All the faces and angles of the one correspond to the faces and angles of the other. Yet, lift them about as we please, we could never fit them together. If we fit the bases together the two will lie on opposite sides, one being below the other. But the dweller in four dimensions of space will fit them together without any trouble. By the mere turning over of one he will convert it into the other without any change whatever in the relative position of its parts. What he could do with the pyramids he could also do with one of us if we allowed him to take hold of us and turn a somersault with us in the fourth dimension. We should then come back into our natural space, but changed as if we were seen in a mirror. Everything on us would be changed from right to left, even the seams in our clothes, and every hair on our head. All this would be done without, during any of the motion, any change having occurred in the positions of the parts of the body.

It is very curious that, in these transcendental speculations, the most rigorous mathematical methods correspond to the most mystical ideas of the Swedenborgian and other forms of religion. Right around us, but in a direction which we cannot conceive any more than the inhabitants of "flat-land" can conceive up and down, there may exist not merely another universe, but any number of universes. All that physical science can say against the supposition is that, even if a fourth dimension exists, there is some law of all the matter with which we are acquainted which prevents any of it from entering that dimension, so that, in our natural condition, it must forever remain unknown to us.

Another possibility in space of four dimensions would be that of turning a hollow sphere, an india-rubber ball, for example, inside out by simple bending without tearing it. To show the motion in our space to which this is analogous, let us take a thin, round sheet of india-rubber, and cut out all the central part, leaving only a narrow ring round the border. Suppose the outer edge of this ring fastened down on a table, while we take hold of the inner edge and stretch it upward and outward over the outer edge until we flatten the whole ring on the table, upside down, with the inner edge now the outer one. This motion would be as inconceivable in "flat-land" as turning the ball inside out is to us.

THE ORGANIZATION OF SCIENTIFIC RESEARCH

The claims of scientific research on the public were never more forcibly urged than in Professor Ray Lankester's recent Romanes Lecture before the University of Oxford. Man is here eloquently pictured as Nature's rebel, who, under conditions where his great superior commands "Thou shalt die," replies "I will live." In pursuance of this determination, civilized man has proceeded so far in his interference with the regular course of Nature that he must either go on and acquire firmer control of the conditions, or perish miserably by the vengeance certain to be inflicted on the half-hearted meddler in great affairs. This rebel by every step forward renders himself liable to greater and greater penalties, and so cannot afford to pause or fail in one single step. One of Nature's most powerful agencies in thwarting his determination to live is found in disease-producing parasites. "Where there is one man of first-rate intelligence now employed in gaining knowledge of this agency, there should be a thousand. It should be as much the purpose of civilized nations to protect their citizens in this respect as it is to provide defence against human aggression."

It was no part of the function of the lecturer to devise a plan for carrying on the great war he proposes to wage. The object of the present article is to contribute some suggestions in this direction; with especial reference to conditions in our own country; and no better text can be found for a discourse on the subject than the preceding quotation. In saying that there should be a thousand investigators of disease where there is now one, I believe that Professor Lankester would be the first to admit that this statement was that of an ideal to be aimed at, rather than of an end to be practically reached. Every careful thinker will agree that to gather a body of men, young or old, supply them with laboratories and microscopes, and tell them to investigate disease, would be much like sending out an army without trained leaders to invade an enemy's country.

There is at least one condition of success in this line which is better fulfilled in our own country than in any other; and that is liberality of support on the part of munificent citizens desirous of so employing their wealth as to promote the public good. Combining this instrumentality with the general public spirit of our people, it must be admitted that, with all the disadvantages under which scientific research among us has hitherto labored, there is still no country to which we can look more hopefully than to our own as the field in which the ideal set forth by Professor Lankester is to be pursued. Some thoughts on the question how scientific research may be most effectively promoted in our own country through organized effort may therefore be of interest. Our first step will be to inquire what general lessons are to be learned from the experience of the past.

The first and most important of these lessons is that research has never reached its highest development except at centres where bodies of men engaged in it have been brought together, and stimulated to action by mutual sympathy and support. We must call to mind that, although the beginnings of modern science were laid by such men as Copernicus, Galileo, Leonardo da Vinci, and Torricelli, before the middle of the seventeenth century, unbroken activity and progress date from the foundations of the Academy of Sciences of Paris and the Royal Society of London at that time. The historic fact that the bringing of men together, and their support by an intelligent and interested community, is the first requirement to be kept in view can easily be explained. Effective research involves so intricate a network of problems and considerations that no one engaged in it can fail to profit by the suggestions of kindred spirits, even if less acquainted with the subject than he is himself. Intelligent discussion suggests new ideas and continually carries the mind to a higher level of thought. We must not regard the typical scientific worker, even of the highest class, as one who, having chosen his special field and met with success in cultivating it, has only to be supplied with the facilities he may be supposed to need in order to continue his work in the most efficient way. What we have to deal with is not a fixed and permanent body of learned men, each knowing all about the field of work in which he is engaged, but a changing and growing class, constantly recruited by beginners at the bottom of the scale, and constantly depleted by the old dropping away at the top. No view of the subject is complete which does not embrace the entire activity of the investigator, from the tyro to the leader. The leader himself, unless engaged in the prosecution of some narrow specialty, can rarely be so completely acquainted with his field as not to need information from others. Without this, he is constantly liable to be repeating what has already been better done than he can do it himself, of following lines which are known to lead to no result, and of adopting methods shown by the experience of others not to be the best. Even the books and published researches to which he must have access may be so voluminous that he cannot find time to completely examine them for himself; or they may be inaccessible. All this will make it clear that, with an occasional exception, the best results of research are not to be expected except at centres where large bodies of men are brought into close personal contact.

In addition to the power and facility acquired by frequent discussion with his fellows, the appreciation and support of an intelligent community, to whom the investigator may, from time to time, make known his thoughts and the results of his work, add a most effective stimulus. The greater the number of men of like minds that can be brought together and the larger the community which interests itself in what they are doing, the more rapid will be the advance and the more effective the work carried on. It is thus that London, with its munificently supported institutions, and Paris and Berlin, with their bodies of investigators supported either by the government or by various foundations, have been for more than three centuries the great centres where we find scientific activity most active and most effective. Looking at this undoubted fact, which has asserted itself through so long a period, and which asserts itself today more strongly than ever, the writer conceives that there can be no question as to one proposition. If we aim at the single object of promoting the advance of knowledge in the most effective way, and making our own country the leading one in research, our efforts should be directed towards bringing together as many scientific workers as possible at a single centre, where they can profit in the highest degree by mutual help, support, and sympathy.

In thus strongly setting forth what must seem an indisputable conclusion, the writer does not deny that there are drawbacks to such a policy, as there are to every policy that can be devised aiming at a good result. Nature offers to society no good that she does not accompany by a greater or less measure of evil The only question is whether the good outweighs the evil. In the present case, the seeming evil, whether real or not, is that of centralization. A policy tending in this direction is held to be contrary to the best interests of science in quarters entitled to so much respect that we must inquire into the soundness of the objection.

It would be idle to discuss so extreme a question as whether we shall take all the best scientific investigators of our country from their several seats of learning and attract them to some one point. We know that this cannot be done, even were it granted that success would be productive of great results. The most that can be done is to choose some existing centre of learning, population, wealth, and influence, and do what we can to foster the growth of science at that centre by attracting thither the greatest possible number of scientific investigators, especially of the younger class, and making it possible for them to pursue their researches in the most effective way. This policy would not result in the slightest harm to any institution or community situated elsewhere. It would not be even like building up a university to outrank all the others of our country; because the functions of the new institution, if such should be founded, would in its relations to the country be radically different from those of a university. Its primary object would not be the education of youth, but the increase of knowledge. So far as the interests of any community or of the world at large are concerned, it is quite indifferent where knowledge may be acquired, because, when once acquired and made public, it is free to the world. The drawbacks suffered by other centres would be no greater than those suffered by our Western cities, because all the great departments of the government are situated at a single distant point. Strong arguments could doubtless be made for locating some of these departments in the Far West, in the Mississippi Valley, or in various cities of the Atlantic coast; but every one knows that any local advantages thus gained would be of no importance compared with the loss of that administrative efficiency which is essential to the whole country.

There is, therefore, no real danger from centralization. The actual danger is rather in the opposite direction; that the sentiment against concentrating research will prove to operate too strongly. There is a feeling that it is rather better to leave every investigator where he chances to be at the moment, a feeling which sometimes finds expression in the apothegm that we cannot transplant a genius. That such a proposition should find acceptance affords a striking example of the readiness of men to accept a euphonious phrase without inquiring whether the facts support the doctrine which it enunciates. The fact is that many, perhaps the majority, of the great scientific investigators of this and of former times have done their best work through being transplanted. As soon as the enlightened monarchs of Europe felt the importance of making their capitals great centres of learning, they began to invite eminent men of other countries to their own. Lagrange was an Italian transplanted to Paris, as a member of the Academy of Sciences, after he had shown his powers in his native country. His great contemporary, Euler, was a Swiss, transplanted first to St. Petersburg, then invited by Frederick the Great to become a member of the Berlin Academy, then again attracted to St. Petersburg. Huyghens was transplanted from his native country to Paris. Agassiz was an exotic, brought among us from Switzerland, whose activity during the generation he passed among us was as great and effective as at any time of his life. On the Continent, outside of France, the most eminent professors in the universities have been and still are brought from distant points. So numerous are the cases of which these are examples that it would be more in accord with the facts to claim that it is only by transplanting a genius that we stimulate him to his best work.

Having shown that the best results can be expected only by bringing into contact as many scientific investigators as possible, the next question which arises is that of their relations to one another. It may be asked whether we shall aim at individualism or collectivism. Shall our ideal be an organized system of directors, professors, associates, assistants, fellows; or shall it be a collection of individual workers, each pursuing his own task in the way he deems best, untrammelled by authority?

The reply to this question is that there is in this special case no antagonism between the two ideas. The most effective organization will aim both at the promotion of individual effort, and at subordination and co-operation. It would be a serious error to formulate any general rule by which all cases should be governed. The experience of the past should be our guide, so far as it applies to present and future conditions; but in availing ourselves of it we must remember that conditions are constantly changing, and must adapt our policy to the problems of the future. In doing this, we shall find that different fields of research require very different policies as regards co-operation and subordination. It will be profitable to point out those special differences, because we shall thereby gain a more luminous insight into the problems which now confront the scientific investigator, and better appreciate their variety, and the necessity of different methods of dealing with them.

At one extreme, we have the field of normative science, work in which is of necessity that of the individual mind alone. This embraces pure mathematics and the methods of science in their widest range. The common interests of science require that these methods shall be worked out and formulated for the guidance of investigators generally, and this work is necessarily that of the individual brain.

At the other extreme, we have the great and growing body of sciences of observation. Through the whole nineteenth century, to say nothing of previous centuries, organizations, and even individuals, have been engaged in recording the innumerable phases of the course of nature, hoping to accumulate material that posterity shall be able to utilize for its benefit. We have observations astronomical, meteorological, magnetic, and social, accumulating in constantly increasing volume, the mass of which is so unmanageable with our present organizations that the question might well arise whether almost the whole of it will not have to be consigned to oblivion. Such a conclusion should not be entertained until we have made a vigorous effort to find what pure metal of value can be extracted from the mass of ore. To do this requires the co-operation of minds of various orders, quite akin in their relations to those necessary in a mine or great manufacturing establishment. Laborers whose duties are in a large measure matters of routine must be guided by the skill of a class higher in quality and smaller in number than their own, and these again by the technical knowledge of leaders in research. Between these extremes we have a great variety of systems of co-operation.

There is another feature of modern research the apprehension of which is necessary to the completeness of our view. A cursory survey of the field of science conveys the impression that it embraces only a constantly increasing number of disconnected specialties, in which each cultivator knows little or nothing of what is being done by others. Measured by its bulk, the published mass of scientific research is increasing in a more than geometrical ratio. Not only do the publications of nearly every scientific society increase in number and volume, but new and vigorous societies are constantly organized to add to the sum total. The stately quartos issued from the presses of the leading academies of Europe are, in most cases, to be counted by hundreds. The Philosophical Transactions of the Royal Society already number about two hundred volumes, and the time when the Memoirs of the French Academy of Sciences shall reach the thousand mark does not belong to the very remote future. Besides such large volumes, these and other societies publish smaller ones in a constantly growing number. In addition to the publications of learned societies, there are journals devoted to each scientific specialty, which seem to propagate their species by subdivision in much the same way as some of the lower orders of animal life. Every new publication of the kind is suggested by the wants of a body of specialists, who require a new medium for their researches and communications. The time has already come when we cannot assume that any specialist is acquainted with all that is being done even in his own line. To keep the run of this may well be beyond his own powers; more he can rarely attempt.

What is the science of the future to do when this huge mass outgrows the space that can be found for it in the libraries, and what are we to say of the value of it all? Are all these scientific researches to be classed as really valuable contributions to knowledge, or have we only a pile in which nuggets of gold are here and there to be sought for? One encouraging answer to such a question is that, taking the interests of the world as a whole, scientific investigation has paid for itself in benefits to humanity a thousand times over, and that all that is known to-day is but an insignificant fraction of what Nature has to show us. Apart from this, another feature of the science of our time demands attention. While we cannot hope that the multiplication of specialties will cease, we find that upon the process of differentiation and subdivision is now being superposed a form of evolution, tending towards the general unity of all the sciences, of which some examples may be pointed out.

Biological science, which a generation ago was supposed to be at the antipodes of exact science, is becoming more and more exact, and is cultivated by methods which are developed and taught by mathematicians. Psychophysics—the study of the operations of the mind by physical apparatus of the same general nature as that used by the chemist and physicist—is now an established branch of research. A natural science which, if any comparisons are possible, may outweigh all others in importance to the race, is the rising one of "eugenics,"—the improvement of the human race by controlling the production of its offspring. No better example of the drawbacks which our country suffers as a seat of science can be given than the fact that the beginning of such a science has been possible only at the seat of a larger body of cultivated men than our land has yet been able to bring together. Generations may elapse before the seed sown by Mr. Francis Galton, from which grew the Eugenic Society, shall bear full fruit in the adoption of those individual efforts and social regulations necessary to the propagation of sound and healthy offspring on the part of the human family. But when this comes about, then indeed will Professor Lankester's "rebel against Nature" find his independence acknowledged by the hitherto merciless despot that has decreed punishment for his treason.

This new branch of science from which so much may be expected is the offshoot of another, the rapid growth of which illustrates the rapid invasion of the most important fields of thought by the methods of exact science. It is only a few years since it was remarked of Professor Karl Pearson's mathematical investigations into the laws of heredity, and the biological questions associated with these laws, that he was working almost alone, because the biologists did not understand his mathematics, while the mathematicians were not interested in his biology. Had he not lived at a great centre of active thought, within the sphere of influence of the two great universities of England, it is quite likely that this condition of isolation would have been his to the end. But, one by one, men were found possessing the skill and interest in the subject necessary to unite in his work, which now has not only a journal of its own, but is growing in a way which, though slow, has all the marks of healthy progress towards an end the importance of which has scarcely dawned upon the public mind.

Admitting that an organized association of investigators is of the first necessity to secure the best results in the scientific work of the future, we meet the question of the conditions and auspices under which they are to be brought together. The first thought to strike us at this point may well be that we have, in our great universities, organizations which include most of the leading men now engaged in scientific research, whose personnel and facilities we should utilize. Admitting, as we all do, that there are already too many universities, and that better work would be done by a consolidation of the smaller ones, a natural conclusion is that the end in view will be best reached through existing organizations. But it would be a great mistake to jump at this conclusion without a careful study of the conditions. The brief argument—there are already too many institutions—instead of having more we should strengthen those we have—should not be accepted without examination. Had it been accepted thirty years ago, there are at least two great American universities of to-day which would not have come into being, the means devoted to their support having been divided among others. These are the Johns Hopkins and the University of Chicago. What would have been gained by applying the argument in these cases? The advantage would have been that, instead of 146 so-called universities which appear to-day in the Annual Report of the Bureau of Education, we should have had only 144. The work of these 144 would have been strengthened by an addition, to their resources, represented by the endowments of Baltimore and Chicago, and sufficient to add perhaps one professor to the staff of each. Would the result have been better than it actually has been? Have we not gained anything by allowing the argument to be forgotten in the cases of these two institutions? I do not believe that any who carefully look at the subject will hesitate in answering this question in the affirmative. The essential point is that the Johns Hopkins University did not merely add one to an already overcrowded list, but that it undertook a mission which none of the others was then adequately carrying out. If it did not plant the university idea in American soil, it at least gave it an impetus which has now made it the dominant one in the higher education of almost every state.

The question whether the country at large would have reaped a greater benefit, had the professors of the University of Chicago, with the appliances they now command, been distributed among fifty or a hundred institutions in every quarter of the land, than it has actually reaped from that university, is one which answers itself. Our two youngest universities have attained success, not because two have thus been added to the number of American institutions of learning, but because they had a special mission, required by the advance of the age, for which existing institutions were inadequate.

The conclusion to which these considerations lead is simple. No new institution is needed to pursue work on traditional lines, guided by traditional ideas. But, if a new idea is to be vigorously prosecuted, then a young and vigorous institution, specially organized to put the idea into effect, is necessary. The project of building up in our midst, at the most appropriate point, an organization of leading scientific investigators, for the single purpose of giving a new impetus to American science and, if possible, elevating the thought of the country and of the world to a higher plane, involves a new idea, which can best be realized by an institution organized for the special purpose. While this purpose is quite in line with that of the leading universities, it goes too far beyond them to admit of its complete attainment through their instrumentality. The first object of a university is the training of the growing individual for the highest duties of life. Additions to the mass of knowledge have not been its principal function, nor even an important function in our own country, until a recent time. The primary object of the proposed institution is the advance of knowledge and the opening up of new lines of thought, which, it may be hoped, are to prove of great import to humanity. It does not follow that the function of teaching shall be wholly foreign to its activities. It must take up the best young men at the point where universities leave them, and train them in the arts of thinking and investigating. But this training will be beyond that which any regular university is carrying out.

In pursuing our theme the question next arises as to the special features of the proposed association. The leading requirement is one that cannot be too highly emphasized. How clearly soever the organizers may have in their minds' eye the end in view, they must recognize the fact that it cannot be attained in a day. In every branch of work which is undertaken, there must be a single leader, and he must be the best that the country, perhaps even the world, can produce. The required man is not to be found without careful inquiry; in many branches he may be unattainable for years. When such is the case, wait patiently till he appears. Prudence requires that the fewest possible risks would be taken, and that no leader should be chosen except one of tried experience and world-wide reputation. Yet we should not leave wholly out of sight the success of the Johns Hopkins University in selecting, at its very foundation, young men who were to prove themselves the leaders of the future. This experience may admit of being repeated, if it be carefully borne in mind that young men of promise are to be avoided and young men of performance only to be considered. The performance need not be striking: ex pede Herculem may be possible; but we must be sure of the soundness of our judgment before accepting our Hercules. This requires a master. Clerk-Maxwell, who never left his native island to visit our shores, is entitled to honor as a promoter of American science for seeing the lion's paw in the early efforts of Rowland, for which the latter was unable to find a medium of publication in his own country. It must also be admitted that the task is more serious now than it was then, because, from the constantly increasing specialization of science, it has become difficult for a specialist in one line to ascertain the soundness of work in another. With all the risks that may be involved in the proceeding, it will be quite possible to select an effective body of leaders, young and old, with whom an institution can begin. The wants of these men will be of the most varied kind. One needs scarcely more than a study and library; another must have small pieces of apparatus which he can perhaps design and make for himself. Another may need apparatus and appliances so expensive that only an institution at least as wealthy as an ordinary university would be able to supply them. The apparatus required by others will be very largely human— assistants of every grade, from university graduates of the highest standing down to routine drudges and day-laborers. Workrooms there must be; but it is hardly probable that buildings and laboratories of a highly specialized character will be required at the outset. The best counsel will be necessary at every step, and in this respect the institution must start from simple beginnings and grow slowly. Leaders must be added one by one, each being judged by those who have preceded him before becoming in his turn a member of the body. As the body grows its members must be kept in personal touch, talk together, pull together, and act together.

The writer submits these views to the great body of his fellow-citizens interested in the promotion of American science with the feeling that, though his conclusions may need amendment in details, they rest upon facts of the past and present which have not received the consideration which they merit. What he most strongly urges is that the whole subject of the most efficient method of promoting research upon a higher plane shall be considered with special reference to conditions in our own country; and that the lessons taught by the history and progress of scientific research in all countries shall be fully weighed and discussed by those most interested in making this form of effort a more important feature of our national life. When this is done, he will feel that his purpose in inviting special consideration to his individual views has been in great measure reached.

CAN WE MAKE IT RAIN

To the uncritical observer the possible achievements of invention and discovery seem boundless. Half a century ago no idea could have appeared more visionary than that of holding communication in a few seconds of time with our fellows in Australia, or having a talk going on viva voce between a man in Washington and another in Boston. The actual attainment of these results has naturally given rise to the belief that the word "impossible" has disappeared from our vocabulary. To every demonstration that a result cannot be reached the answer is, Did not one Lardner, some sixty years ago, demonstrate that a steamship could not cross the Atlantic? If we say that for every actual discovery there are a thousand visionary projects, we are told that, after all, any given project may be the one out of the thousand.

In a certain way these hopeful anticipations are justified. We cannot set any limit either to the discovery of new laws of nature or to the ingenious combination of devices to attain results which now look impossible. The science of to-day suggests a boundless field of possibilities. It demonstrates that the heat which the sun radiates upon the earth in a single day would suffice to drive all the steamships now on the ocean and run all the machinery on the land for a thousand years. The only difficulty is how to concentrate and utilize this wasted energy. From the stand-point of exact science aerial navigation is a very simple matter. We have only to find the proper combination of such elements as weight, power, and mechanical force. Whenever Mr. Maxim can make an engine strong and light enough, and sails large, strong, and light enough, and devise the machinery required to connect the sails and engine, he will fly. Science has nothing but encouraging words for his project, so far as general principles are concerned. Such being the case, I am not going to maintain that we can never make it rain.

But I do maintain two propositions. If we are ever going to make it rain, or produce any other result hitherto unattainable, we must employ adequate means. And if any proposed means or agency is already familiar to science, we may be able to decide beforehand whether it is adequate. Let us grant that out of a thousand seemingly visionary projects one is really sound. Must we try the entire thousand to find the one? By no means. The chances are that nine hundred of them will involve no agency that is not already fully understood, and may, therefore, be set aside without even being tried. To this class belongs the project of producing rain by sound. As I write, the daily journals are announcing the brilliant success of experiments in this direction; yet I unhesitatingly maintain that sound cannot make rain, and propose to adduce all necessary proof of my thesis. The nature of sound is fully understood, and so are the conditions under which the aqueous vapor in the atmosphere may be condensed. Let us see how the case stands.

A room of average size, at ordinary temperature and under usual conditions, contains about a quart of water in the form of invisible vapor. The whole atmosphere is impregnated with vapor in about the same proportion. We must, however, distinguish between this invisible vapor and the clouds or other visible masses to which the same term is often applied. The distinction may be very clearly seen by watching the steam coming from the spout of a boiling kettle. Immediately at the spout the escaping steam is transparent and invisible; an inch or two away a white cloud is formed, which we commonly call steam, and which is seen belching out to a distance of one or more feet, and perhaps filling a considerable space around the kettle; at a still greater distance this cloud gradually disappears. Properly speaking, the visible cloud is not vapor or steam at all, but minute particles or drops of water in a liquid state. The transparent vapor at the mouth of the kettle is the true vapor of water, which is condensed into liquid drops by cooling; but after being diffused through the air these drops evaporate and again become true vapor. Clouds, then, are not formed of true vapor, but consist of impalpable particles of liquid water floating or suspended in the air.

But we all know that clouds do not always fall as rain. In order that rain may fall the impalpable particles of water which form the cloud must collect into sensible drops large enough to fall to the earth. Two steps are therefore necessary to the formation of rain: the transparent aqueous vapor in the air must be condensed into clouds, and the material of the clouds must agglomerate into raindrops.

No physical fact is better established than that, under the conditions which prevail in the atmosphere, the aqueous vapor of the air cannot be condensed into clouds except by cooling. It is true that in our laboratories it can be condensed by compression. But, for reasons which I need not explain, condensation by compression cannot take place in the air. The cooling which results in the formation of clouds and rain may come in two ways. Rains which last for several hours or days are generally produced by the intermixture of currents of air of different temperatures. A current of cold air meeting a current of warm, moist air in its course may condense a considerable portion of the moisture into clouds and rain, and this condensation will go on as long as the currents continue to meet. In a hot spring day a mass of air which has been warmed by the sun, and moistened by evaporation near the surface of the earth, may rise up and cool by expansion to near the freezing-point. The resulting condensation of the moisture may then produce a shower or thunder-squall. But the formation of clouds in a clear sky without motion of the air or change in the temperature of the vapor is simply impossible. We know by abundant experiments that a mass of true aqueous vapor will never condense into clouds or drops so long as its temperature and the pressure of the air upon it remain unchanged.

Now let us consider sound as an agent for changing the state of things in the air. It is one of the commonest and simplest agencies in the world, which we can experiment upon without difficulty. It is purely mechanical in its action. When a bomb explodes, a certain quantity of gas, say five or six cubic yards, is suddenly produced. It pushes aside and compresses the surrounding air in all directions, and this motion and compression are transmitted from one portion of the air to another. The amount of motion diminishes as the square of the distance; a simple calculation shows that at a quarter of a mile from the point of explosion it would not be one ten-thousandth of an inch. The condensation is only momentary; it may last the hundredth or the thousandth of a second, according to the suddenness and violence of the explosion; then elasticity restores the air to its original condition and everything is just as it was before the explosion. A thousand detonations can produce no more effect upon the air, or upon the watery vapor in it, than a thousand rebounds of a small boy's rubber ball would produce upon a stonewall. So far as the compression of the air could produce even a momentary effect, it would be to prevent rather than to cause condensation of its vapor, because it is productive of heat, which produces evaporation, not condensation.

The popular notion that sound may produce rain is founded principally upon the supposed fact that great battles have been followed by heavy rains. This notion, I believe, is not confirmed by statistics; but, whether it is or not, we can say with confidence that it was not the sound of the cannon that produced the rain. That sound as a physical factor is quite insignificant would be evident were it not for our fallacious way of measuring it. The human ear is an instrument of wonderful delicacy, and when its tympanum is agitated by a sound we call it a "concussion" when, in fact, all that takes place is a sudden motion back and forth of a tenth, a hundredth, or a thousandth of an inch, accompanied by a slight momentary condensation. After these motions are completed the air is exactly in the same condition as it was before; it is neither hotter nor colder; no current has been produced, no moisture added.

If the reader is not satisfied with this explanation, he can try a very simple experiment which ought to be conclusive. If he will explode a grain of dynamite, the concussion within a foot of the point of explosion will be greater than that which can be produced by the most powerful bomb at a distance of a quarter of a mile. In fact, if the latter can condense vapor a quarter of a mile away, then anybody can condense vapor in a room by slapping his hands. Let us, therefore, go to work slapping our hands, and see how long we must continue before a cloud begins to form.

What we have just said applies principally to the condensation of invisible vapor. It may be asked whether, if clouds are already formed, something may not be done to accelerate their condensation into raindrops large enough to fall to the ground. This also may be the subject of experiment. Let us stand in the steam escaping from a kettle and slap our hands. We shall see whether the steam condenses into drops. I am sure the experiment will be a failure; and no other conclusion is possible than that the production of rain by sound or explosions is out of the question.

It must, however, be added that the laws under which the impalpable particles of water in clouds agglomerate into drops of rain are not yet understood, and that opinions differ on this subject. Experiments to decide the question are needed, and it is to be hoped that the Weather Bureau will undertake them. For anything we know to the contrary, the agglomeration may be facilitated by smoke in the air. If it be really true that rains have been produced by great battles, we may say with confidence that they were produced by the smoke from the burning powder rising into the clouds and forming nuclei for the agglomeration into drops, and not by the mere explosion. If this be the case, if it was the smoke and not the sound that brought the rain, then by burning gunpowder and dynamite we are acting much like Charles Lamb's Chinamen who practised the burning of their houses for several centuries before finding out that there was any cheaper way of securing the coveted delicacy of roast pig.

But how, it may be asked, shall we deal with the fact that Mr. Dyrenforth's recent explosions of bombs under a clear sky in Texas were followed in a few hours, or a day or two, by rains in a region where rain was almost unknown? I know too little about the fact, if such it be, to do more than ask questions about it suggested by well-known scientific truths. If there is any scientific result which we can accept with confidence, it is that ten seconds after the sound of the last bomb died away, silence resumed her sway. From that moment everything in the air—humidity, temperature, pressure, and motion—was exactly the same as if no bomb had been fired. Now, what went on during the hours that elapsed between the sound of the last bomb and the falling of the first drop of rain? Did the aqueous vapor already in the surrounding air slowly condense into clouds and raindrops in defiance of physical laws? If not, the hours must have been occupied by the passage of a mass of thousands of cubic miles of warm, moist air coming from some other region to which the sound could not have extended. Or was Jupiter Pluvius awakened by the sound after two thousand years of slumber, and did the laws of nature become silent at his command? When we transcend what is scientifically possible, all suppositions are admissible; and we leave the reader to take his choice between these and any others he may choose to invent.

One word in justification of the confidence with which I have cited established physical laws. It is very generally supposed that most great advances in applied science are made by rejecting or disproving the results reached by one's predecessors. Nothing could be farther from the truth. As Huxley has truly said, the army of science has never retreated from a position once gained. Men like Ohm and Maxwell have reduced electricity to a mathematical science, and it is by accepting, mastering, and applying the laws of electric currents which they discovered and expounded that the electric light, electric railway, and all other applications of electricity have been developed. It is by applying and utilizing the laws of heat, force, and vapor laid down by such men as Carnot and Regnault that we now cross the Atlantic in six days. These same laws govern the condensation of vapor in the atmosphere; and I say with confidence that if we ever do learn to make it rain, it will be by accepting and applying them, and not by ignoring or trying to repeal them.

How much the indisposition of our government to secure expert scientific evidence may cost it is strikingly shown by a recent example. It expended several million dollars on a tunnel and water-works for the city of Washington, and then abandoned the whole work. Had the project been submitted to a commission of geologists, the fact that the rock-bed under the District of Columbia would not stand the continued action of water would have been immediately reported, and all the money expended would have been saved. The fact is that there is very little to excite popular interest in the advance of exact science. Investigators are generally quiet, unimpressive men, rather diffident, and wholly wanting in the art of interesting the public in their work. It is safe to say that neither Lavoisier, Galvani, Ohm, Regnault, nor Maxwell could have gotten the smallest appropriation through Congress to help make discoveries which are now the pride of our century. They all dealt in facts and conclusions quite devoid of that grandeur which renders so captivating the project of attacking the rains in their aerial stronghold with dynamite bombs.

THE ASTRONOMICAL EPHEMERIS AND THE
NAUTICAL ALMANAC

[Footnote: Read before the U S Naval Institute, January 10, 1879.]
Although the Nautical Almanacs of the world, at the present time, are of comparatively recent origin, they have grown from small beginnings, the tracing of which is not unlike that of the origin of species by the naturalist of the present day. Notwithstanding its familiar name, it has always been designed rather for astronomical than for nautical purposes. Such a publication would have been of no use to the navigator before he had instruments with which to measure the altitudes of the heavenly bodies. The earlier navigators seldom ventured out of sight of land, and during the night they are said to have steered by the "Cynosure" or constellation of the Great Bear, a practice which has brought the name of the constellation into our language of the present day to designate an object on which all eyes are intently fixed. This constellation was a little nearer the pole in former ages than at the present time; still its distance was always so great that its use as a mark of the northern point of the horizon does not inspire us with great respect for the accuracy with which the ancient navigators sought to shape their course.

The Nautical Almanac of the present day had its origin in the Astronomical Ephemerides called forth by the needs of predictions of celestial motions both on the part of the astronomer and the citizen. So long as astrology had a firm hold on the minds of men, the positions of the planets were looked to with great interest. The theories of Ptolemy, although founded on a radically false system, nevertheless sufficed to predict the position of the sun, moon, and planets, with all the accuracy necessary for the purposes of the daily life of the ancients or the sentences of their astrologers. Indeed, if his tables were carried down to the present time, the positions of the heavenly bodies would be so few degrees in error that their recognition would be very easy. The times of most of the eclipses would be predicted within a few hours, and the conjunctions of the planets within a few days. Thus it was possible for the astronomers of the Middle Ages to prepare for their own use, and that of the people, certain rude predictions respecting the courses of the sun and moon and the aspect of the heavens, which served the purpose of daily life and perhaps lessened the confusion arising from their complicated calendars. In the signs of the zodiac and the different effects which follow from the sun and moon passing from sign to sign, still found in our farmers' almanacs, we have the dying traces of these ancient ephemerides.

The great Kepler was obliged to print an astrological almanac in virtue of his position as astronomer of the court of the King of Austria. But, notwithstanding the popular belief that astronomy had its origin in astrology, the astronomical writings of all ages seem to show that the astronomers proper never had any belief in astrology. To Kepler himself the necessity for preparing this almanac was a humiliation to which he submitted only through the pressure of poverty. Subsequent ephemerides were prepared with more practical objects. They gave the longitudes of the planets, the position of the sun, the time of rising and setting, the prediction of eclipses, etc.

They have, of course, gradually increased in accuracy as the tables of the celestial motions were improved from time to time. At first they were not regular, annual publications, issued by governments, as at the present time, but the works of individual astronomers who issued their ephemerides for several years in advance, at irregular intervals. One man might issue one, two, or half a dozen such volumes, as a private work, for the benefit of his fellows, and each might cover as many years as he thought proper.

The first publication of this sort, which I have in my possession, is the Ephemerides of Manfredi, of Bonn, computed for the years 1715 to 1725, in two volumes.

Of the regular annual ephemerides the earliest, so far as I am aware, is the Connaissance des Temps or French Nautical Almanac. The first issue was in the year 1679, by Picard, and it has been continued without interruption to the present time. Its early numbers were, of course, very small, and meagre in their details. They were issued by the astronomers of the French Academy of Sciences, under the combined auspices of the academy and the government. They included not merely predictions from the tables, but also astronomical observations made at the Paris Observatory or elsewhere. When the Bureau of Longitudes was created in 1795, the preparation of the work was intrusted to it, and has remained in its charge until the present time. As it is the oldest, so, in respect at least to number of pages, it is the largest ephemeris of the present time. The astronomical portion of the volume for 1879 fills more than seven hundred pages, while the table of geographical positions, which has always been a feature of the work, contains nearly one hundred pages more.

The first issue of the British Nautical Almanac was that for the year 1767 and appeared in 1766. It differs from the French Almanac in owing its origin entirely to the needs of navigation. The British nation, as the leading maritime power of the world, was naturally interested in the discovery of a method by which the longitude could be found at sea. As most of my hearers are probably aware, there was, for many years, a standing offer by the British government, of ten thousand pounds for the discovery of a practical and sufficiently accurate method of attaining this object. If I am rightly informed, the requirement was that a ship should be able to determine the Greenwich time within two minutes, after being six months at sea. When the office of Astronomer Royal was established in 1765, the duty of the incumbent was declared to be "to apply himself with the most exact care and diligence to the rectifying the Tables of the Motions of the Heavens, and the places of the Fixed Stars in order to find out the so much desired Longitude at Sea for the perfecting the Art of Navigation."

About the middle of the last century the lunar tables were so far improved that Dr. Maskelyne considered them available for attaining this long-wished-for object. The method which I think was then, for the first time, proposed was the now familiar one of lunar distances. Several trials of the method were made by accomplished gentlemen who considered that nothing was wanting to make it practical at sea but a Nautical Ephemeris. The tables of the moon, necessary for the purpose, were prepared by Tobias Mayer, of Gottingen, and the regular annual issue of the work was commenced in 1766, as already stated. Of the reward which had been offered, three thousand pounds were paid to the widow of Mayer, and three thousand pounds to the celebrated mathematician Euler for having invented the methods used by Mayer in the construction of his tables. The issue of the Nautical Ephemeris was intrusted to Dr. Maskelyne. Like other publications of this sort this ephemeris has gradually increased in volume. During the first sixty or seventy years the data were extremely meagre, including only such as were considered necessary for the determination of positions.

In 1830 the subject of improving the Nautical Almanac was referred by the Lord Commissioners of the Admiralty to a committee of the Astronomical Society of London. A subcommittee, including eleven of the most distinguished astronomers and one scientific navigator, made an exhaustive report, recommending a radical rearrangement and improvement of the work. The recommendations of this committee were first carried into effect in the Nautical Almanac for the year 1834. The arrangement of the

Navigator's Ephemeris then devised has been continued in the British Almanac to the present time.

A good deal of matter has been added to the British Almanac during the forty years and upwards which have elapsed, but it has been worked in rather by using smaller type and closer printing than by increasing the number of pages. The almanac for 1834 contains five hundred and seventeen pages and that for 1880 five hundred and nineteen pages. The general aspect of the page is now somewhat crowded, yet, considering the quantity of figures on each page the arrangement is marvellously clear and legible.

The Spanish "Almanaque Nautico" has been issued since the beginning of the century. Like its fellows it has been gradually enlarged and improved, in recent times, and is now of about the same number of pages with the British and American almanacs. As a rule there is less matter on a page, so that the data actually given are not so complete as in some other publications.

In Germany two distinct publications of this class are issued, the one purely astronomical, the other purely nautical.

The astronomical publication has been issued for more than a century under the title of "Berliner Astronomisches Jahrbuch." It is intended principally for the theoretical astronomer, and in respect to matter necessary to the determinations of positions on the earth it is rather meagre. It is issued by the Berlin Observatory, at the expense of the government.

The companion of this work, intended for the use of the German marine, is the "Nautisches Jahrbuch," prepared and issued under the direction of the minister of commerce and public works. It is copied largely from the British Nautical Almanac, and in respect to arrangement and data is similar to our American Nautical Almanac, prepared for the use of navigators, giving, however, more matter, but in a less convenient form. The right ascension and declination of the moon are given for every three hours instead of for every hour; one page of each month is devoted to eclipses of Jupiter's satellites, phenomena which we never consider necessary in the nautical portion of our own almanac. At the end of the work the apparent positions of seventy or eighty of the brightest stars are given for every ten days, while it is considered that our own navigators will be satisfied with the mean places for the beginning of the year. At the end is a collection of tables which I doubt whether any other than a German navigator would ever use. Whether they use them or not I am not prepared to say.

The preceding are the principal astronomical and nautical ephemerides of the world, but there are a number of minor publications, of the same class, of which I cannot pretend to give a complete list. Among them is the Portuguese Astronomical Ephemeris for the meridian of the University of Coimbra, prepared for Portuguese navigators. I do not know whether the

Portuguese navigators really reckon their longitudes from this point: if they do the practice must be attended with more or less confusion. All the matter is given by months, as in the solar and lunar ephemeris of our own and the British Almanac. For the sun we have its longitude, right ascension, and declination, all expressed in arc and not in time. The equation of time and the sidereal time of mean noon complete the ephemeris proper. The positions of the principal planets are given in no case oftener than for every third day. The longitude and latitude of the moon are given for noon and midnight. One feature not found in any other almanac is the time at which the moon enters each of the signs of the zodiac. It may be supposed that this information is designed rather for the benefit of the Portuguese landsman than of the navigator. The right ascensions and declinations of the moon and the lunar distances are also given for intervals of twelve hours. Only the last page gives the eclipses of the satellites of Jupiter. The Fixed Stars are wholly omitted.

An old ephemeris, and one well known in astronomy is that published by the Observatory of Milan, Italy, which has lately entered upon the second century of its existence. Its data are extremely meagre and of no interest whatever to the navigator. The greater part of the volume is taken up with observations at the Milan Observatory.

Since taking charge of the American Ephemeris I have endeavored to ascertain what nautical almanacs are actually used by the principal maritime nations of Europe. I have been able to obtain none except those above mentioned. As a general rule I think the British Nautical Almanac is used by all the northern nations, as already indicated. The German Nautical Jahrbuch is principally a reprint from the British. The Swedish navigators, being all well acquainted with the English language, use the British Almanac without change. The Russian government, however, prints an explanation of the various terms in the language of their own people and binds it in at the end of the British Almanac. This explanation includes translations of the principal terms used in the heading of pages, such as the names of the months and days, the different planets, constellations, and fixed stars, and the phenomena of angle and time. They have even an index of their own in which the titles of the different articles are given in Russian. This explanation occupies, in all, seventy-five pages—more than double that taken up by the original explanation.

One of the first considerations which strikes us in comparing these multitudinous publications is the confusion which must arise from the use of so many meridians. If each of these southern nations, the Spanish and Portuguese for instance, actually use a meridian of their own, the practice must lead to great confusion. If their navigators do not do so but refer their longitudes to the meridian of Greenwich, then their almanacs must be as good as useless. They would find it far better to buy an ephemeris referred to the meridian of Greenwich than to attempt to use their own The northern nations, I think, have all begun to refer to the meridian of Greenwich, and the same thing is happily true of our own marine. We may, therefore, hope that all commercial nations will, before long, refer their longitudes to one and the same meridian, and the resulting confusion be thus avoided.

The preparation of the American Ephemeris and Nautical Almanac was commenced in 1849, under the superintendence of the late Rear-Admiral, then Lieutenant, Charles Henry Davis. The first volume to be issued was that for the year 1855. Both in the preparation of that work and in the connected work of mapping the country, the question of the meridian to be adopted was one of the first importance, and received great attention from Admiral Davis, who made an able report on the subject. Our situation was in some respects peculiar, owing to the great distance which separated us from Europe and the uncertainty of the exact difference of longitude between the two continents. It was hardly practicable to refer longitudes in our own country to any European meridian. The attempt to do so would involve continual changes as the transatlantic longitude was from time to time corrected. On the other hand, in order to avoid confusion in navigation, it was essential that our navigators should continue to reckon from the meridian of Greenwich. The trouble arising from uncertainty of the exact longitude does not affect the navigator, because, for his purpose, astronomical precision is not necessary.

The wisest solution was probably that embodied in the act of Congress, approved September 28, 1850, on the recommendation of Lieutenant Davis, if I mistake not. "The meridian of the Observatory at Washington shall be adopted and used as the American meridian for all astronomical purposes, and the meridian of Greenwich shall be adopted for all nautical purposes." The execution of this law necessarily involves the question, "What shall be considered astronomical and what nautical purposes?" Whether it was from the difficulty of deciding this question, or from nobody's remembering the law, the latter has been practically a dead letter. Surely, if there is any region of the globe which the law intended should be referred to the meridian of Washington, it is the interior of our own country. Yet, notwithstanding the law, all acts of Congress relating to the territories have, so far as I know, referred everything to the meridian of Greenwich and not to that of Washington. Even the maps issued by our various surveys are referred to the same transatlantic meridian. The absurdity culminated in a local map of the city of Washington and the District of Columbia, issued by private parties, in 1861, in which we find even the meridians passing through the city of Washington referred to a supposed Greenwich.

This practice has led to a confusion which may not be evident at first sight, but which is so great and permanent that it may be worth explaining. If, indeed, we could actually refer all our longitudes to an accurate meridian of Greenwich in the first place; if, for instance, any western region could be at once connected by telegraph with the Greenwich Observatory, and thus exchange longitude signals night after night, no trouble or confusion would arise from referring to the meridian of Greenwich. But this, practically, cannot be done. All our interior longitudes have been and are determined differentially by comparison with some point in this country. One of the most frequent points of reference used this way has been the Cambridge Observatory. Suppose, then, a surveyor at Omaha makes a telegraphic longitude determination between that point and the Cambridge Observatory. Since he wants his longitude reduced to Greenwich, he finds some supposed longitude of the Cambridge Observatory from Greenwich and adds that to his own longitude. Thus, what he gives is a longitude actually determined, plus an assumed longitude of Cambridge, and, unless the assumed longitude of Cambridge is distinctly marked on his maps, we may not know what it is.

After a while a second party determines the longitude of Ogden from Cambridge. In the mean time, the longitude of Cambridge from Greenwich has been corrected, and we have a longitude of Ogden which will be discordant with that of Omaha, owing to the change in the longitude of Cambridge. A third party determines the longitudes of, let us suppose, St. Louis from Washington, he adds the assumed longitudes of Washington from Greenwich which may not agree with either of the longitudes of Cambridge and gets his longitude. Thus we have a series of results for our western longitude all nominally referred to the meridian of Greenwich, but actually referred to a confused collection of meridians, nobody knows what. If the law had only provided that the longitude of Washington from Greenwich should be invariably fixed at a certain quantity, say 77 degrees 3', this confusion would not have arisen. It is true that the longitude thus established by law might not have been perfectly correct, but this would not cause any trouble nor confusion. Our longitude would have been simply referred to a certain assumed Greenwich, the small error of which would have been of no importance to the navigator or astronomer. It would have differed from the present system only in that the assumed Greenwich would have been invariable instead of dancing about from time to time as it has done under the present system. You understand that when the astronomer, in computing an interior longitude, supposes that of Cambridge from Greenwich to be a certain definite amount, say 4h 44m 30s, what he actually does is to count from a meridian just that far east of Cambridge. When he changes the assumed longitude of Cambridge he counts from a meridian farther east or farther west of his former one: in other words, he always counts from an assumed Greenwich, which changes its position from time to time, relative to our own country.

Having two meridians to look after, the form of the American Ephemeris, to be best adapted to the wants both of navigators and astronomers was necessarily peculiar. Had our navigators referred their longitudes to any meridian of our own country the arrangement of the work need not have differed materially from that of foreign ones. But being referred to a meridian far outside our limits and at the same time designed for use within those limits, it was necessary to make a division of the matter. Accordingly, the American Ephemeris has always been divided into two parts: the first for the use of navigators, referred to the meridian of Greenwich, the second for that of astronomers, referred to the meridian of Washington. The division of the matter without serious duplication is more easy than might at first be imagined. In explaining it, I will take the ephemeris as it now is, with the small changes which have been made from time to time.

One of the purposes of any ephemeris, and especially of that of the navigators, is to give the position of the heavenly bodies at equidistant intervals of time, usually one day. Since it is noon at some point of the earth all the time, it follows that such an ephemeris will always be referred to noon at some meridian. What meridian this shall be is purely a practical question, to be determined by convenience and custom. Greenwich noon, being that necessarily used by the navigator, is adopted as the standard, but we must not conclude that the ephemeris for Greenwich noon is referred to the meridian of Greenwich in the sense that we refer a longitude to that meridian. Greenwich noon is 18h 51m 48s, Washington mean time; so the ephemeris which gives data for every Greenwich noon may be considered as referred to the meridian of Washington giving the data for 17h 51m 48s, Washington time, every day. The rule adopted, therefore, is to have all the ephemerides which refer to absolute time, without any reference to a meridian, given for Greenwich noon, unless there may be some special reason to the contrary. For the needs of the navigator and the theoretical astronomer these are the most convenient epochs.

Another part of the ephemeris gives the position of the heavenly bodies, not at equidistant intervals, but at transit over some meridian. For this purpose the meridian of Washington is chosen for obvious reasons. The astronomical part of our ephemeris, therefore, gives the positions of the principal fixed stars, the sun, moon, and all the larger planets at the moment of transit over our own meridian.

The third class of data in the ephemeris comprises phenomena to be predicted and observed. Such are eclipses of the sun and moon, occultations of fixed stars by the moon, and eclipses of Jupiter's satellites. These phenomena are all given in Washington mean time as being most convenient for observers in our own country. There is a partial exception, however, in the case of eclipses of the sun and moon. The former are rather for the world in general than for our own country, and it was found difficult to arrange them to be referred to the meridian of Washington without having the maps referred to the same meridian. Since, however, the meridian of Greenwich is most convenient outside of our own territory, and since but a small portion of the eclipses are visible within it, it is much the best to have the eclipses referred entirely to the meridian of Greenwich. I am the more ready to adopt this change because when the eclipses are to be computed for our own country the change of meridians will be very readily understood by those who make the computation.

It may be interesting to say something of the tables and theories from which the astronomical ephemerides are computed. To understand them completely it is necessary to trace them to their origin. The problem of calculating the motions of the heavenly bodies and the changes in the aspect of the celestial sphere was one of the first with which the students of astronomy were occupied. Indeed, in ancient times, the only astronomical problems which could be attacked were of this class, for the simple reason that without the telescope and other instruments of research it was impossible to form any idea of the physical constitution of the heavenly bodies. To the ancients the stars and planets were simply points or surfaces in motion. They might have guessed that they were globes like that on which we live, but they were unable to form any theory of the nature of these globes. Thus, in The Almagest of Ptolemy, the most complete treatise on the ancient astronomy which we possess, we find the motions of all the heavenly bodies carefully investigated and tables given for the convenient computation of their positions. Crude and imperfect though these tables may be, they were the beginnings from which those now in use have arisen.

No radical change was made in the general principles on which these theories and tables were constructed until the true system of the world was propounded by Copernicus. On this system the apparent motion of each planet in the epicycle was represented by a motion of the earth around the sun, and the problem of correcting the position of the planet on account of the epicycle was reduced to finding its geocentric from its heliocentric position. This was the greatest step ever taken in theoretical astronomy, yet it was but a single step. So far as the materials were concerned and the mode of representing the planetary motions, no other radical advance was made by Copernicus. Indeed, it is remarkable that he introduced an epicycle which was not considered necessary by Ptolemy in order to represent the inequalities in the motions of the planets around the sun.

The next great advance made in the theory of the planetary motion was the discovery by Kepler of the celebrated laws which bear his name. When it was established that each planet moved in an ellipse having the sun in one focus it became possible to form tables of the motions of the heavenly bodies much more accurate than had before been known. Such tables were published by Kepler in 1632, under the name of Rudolphine Tables, in memory of his patron, the Emperor Rudolph. But the laws of Kepler took no account of the action of the planets on one another. It is well known that if each planet moved only under the influence of the gravitating force of the sun its motion would accord rigorously with the laws of Kepler, and the problems of theoretical astronomy would be greatly simplified. When, therefore, the results of Kepler's laws were compared with ancient and modern observations it was found that they were not exactly represented by the theory. It was evident that the elliptic orbits of the planets were subject to change, but it was entirely beyond the power of investigation, at that time, to assign any cause for such changes. Notwithstanding the simplicity of the causes which we now know to produce them, they are in form extremely complex. Without the knowledge of the theory of gravitation it would be entirely out of the question to form any tables of the planetary motions which would at all satisfy our modern astronomers.

When the theory of universal gravitation was propounded by Newton he showed that a planet subjected only to the gravitation of a central body, like the sun, would move in exact accordance with Kepler's laws. But by his theory the planets must attract one another and these attractions must cause the motions of each to deviate slightly from the laws in question. Since such deviations were actually observed it was very natural to conclude that they were due to this cause, but how shall we prove it? To do this with all the rigor required in a mathematical investigation it is necessary to calculate the effect of the mutual action of the planets in changing their orbits. This calculation must be made with such precision that there shall be no doubt respecting the results of the theory. Then its results must be compared with the best observations. If the slightest outstanding difference is established there is something wrong and the requirements of astronomical science are not satisfied. The complete solution of this problem was entirely beyond the power of Newton. When his methods of research were used he was indeed able to show that the mutual action of the planets would produce deviations in their motions of the same general nature with those observed, but he was not able to calculate these deviations with numerical exactness. His most successful attempt in this direction was perhaps made in the case of the moon. He showed that the sun's disturbing force on this body would produce several inequalities the existence of which had been established by observation, and he was also able to give a rough estimate of their amount, but this was as far as his method could go. A great improvement had to be made, and this was effected not by English, but by continental mathematicians.

The latter saw, clearly, that it was impossible to effect the required solution by the geometrical mode of reasoning employed by Newton. The problem, as it presented itself to their minds, was to find algebraic expressions for the positions of the planets at any time. The latitude, longitude, and radius-vector of each planet are constantly varying, but they each have a determined value at each moment of time. They may therefore be regarded as functions of the time, and the problem was to express these functions by algebraic formulae. These algebraic expressions would contain, besides the time, the elements of the planetary orbits to be derived from observation. The time which we may suppose to be represented algebraically by the symbol t, would remain as an unknown quantity to the end. What the mathematician sought to do was to present the astronomer with a series of algebraic expressions containing t as an indeterminate quantity, and so, by simply substituting for t any year and fraction of a year whatever—1600, 1700, 1800, for example, the result would give the latitude, longitude, or radius-vector of a planet.

The problem as thus presented was one of the most difficult we can perceive of, but the difficulty was only an incentive to attacking it with all the greater energy. So long as the motion was supposed purely elliptical, so long as the action of the planets was neglected, the problem was a simple one, requiring for its solution only the analytic geometry of the ellipse. The real difficulties commenced when the mutual action of the planets was taken into account. It is, of course, out of the question to give any technical description or analysis of the processes which have been invented for solving the problem; but a brief historical sketch may not be out of place. A complete and rigorous solution of the problem is out of the question—that is, it is impossible by any known method to form an algebraic expression for the co-ordinates of a planet which shall be absolutely exact in a mathematical sense. In whatever way we go to work the expression comes out in the form of an infinite series of terms, each term being, on the whole, a little smaller as we increase the number. So, by increasing the number of these various terms, we can approach nearer and nearer to a mathematical exactness, but can never reach it. The mathematician and astronomer have to be satisfied when they have carried the solution so far that the neglected quantities are entirely beyond the powers of observation.

Mathematicians have worked upon the problem in its various phases for nearly two centuries, and many improvements in detail have, from time to time, been made, but no general method, applicable to all cases, has been devised. One plan is to be used in treating the motion of the moon, another for the interior planets, another for Jupiter and Saturn, another for the minor planets, and so on. Under these circumstances it will not surprise you to learn that our tables of the celestial motions do not, in general, correspond in accuracy to the present state of practical astronomy. There is no authority and no office in the world whose duty it is to look after the preparations of the formulae I have described. The work of computing them has been almost entirely left to individual mathematicians whose taste lay in that direction, and who have sometimes devoted the greater part of their lives to calculations on a single part of the work. As a striking instance of this, the last great work on the Motion of the Moon, that of Delaunay, of Paris, involved some fifteen years of continuous hard labor.

Hansen, of Germany, who died five years ago, devoted almost his whole life to investigations of this class and to the development of new methods of computation. His tables of the moon are those now used for predicting the places of the moon in all the ephemerides of the world.

The only successful attempt to prepare systematic tables for all the large planets is that completed by Le Verrier just before his death; but he used only a small fraction of the material at his disposal, and did not employ the modern methods, confining himself wholly to those invented by his countrymen about the beginning of the present century. For him Jacobi and Hansen had lived in vain.

The great difficulty which besets the subject arises from the fact that mathematical processes alone will not give us the position of a planet, there being seven unknown quantities for each planet which must be determined by observations. A planet, for instance, may move in any ellipse whatever, having the sun in one focus, and it is impossible to tell what ellipse it is, except from observation. The mean motion of a planet, or its period of revolution, can only be determined by a long series of observations, greater accuracy being obtained the longer the observations are continued. Before the time of Bradley, who commenced work at the Greenwich Observatory about 1750, the observations were so far from accurate that they are now of no use whatever, unless in exceptional cases. Even Bradley's observations are in many cases far less accurate than those made now. In consequence, we have not heretofore had a sufficiently extended series of observations to form an entirely satisfactory theory of the celestial motions.

As a consequence of the several difficulties and drawbacks, when the computation of our ephemeris was started, in the year 1849, there were no tables which could be regarded as really satisfactory in use. In the British Nautical Almanac the places of the moon were derived from the tables of Burckhardt published in the year 1812. You will understand, in a case like this, no observations subsequent to the issue of the tables are made use of; the place of the moon of any day, hour, and minute of Greenwich time, mean time, was precisely what Burckhardt would have computed nearly a half a century before. Of the tables of the larger planets the latest were those of Bouvard, published in 1812, while the places of Venus were from tables published by Lindenau in 1810. Of course such tables did not possess astronomical accuracy. At that time, in the case of the moon, completely new tables were constructed from the results reached by Professor Airy in his reduction of the Greenwich observations of the moon from 1750 to 1830. These were constructed under the direction of Professor Pierce and represented the places of the moon with far greater accuracy than the older tables of Burckhardt. For the larger planets corrections were applied to the older tables to make them more nearly represent observations before new ones were constructed. These corrections, however, have not proved satisfactory, not being founded on sufficiently thorough investigations. Indeed, the operation of correcting tables by observation, as we would correct the dead-reckoning of a ship, is a makeshift, the result of which must always be somewhat uncertain, and it tends to destroy that unity which is an essential element of the astronomical ephemeris designed for permanent future use. The result of introducing them, while no doubt an improvement on the old tables, has not been all that should be desired. The general lack of unity in the tables hitherto employed is such that I can only state what has been done by mentioning each planet in detail.

For Mercury, new tables were constructed by Professor Winlock, from formulae published by Le Verrier in 1846. These tables have, however, been deviating from the true motion of the planet, owing to the motion of the perihelion of Mercury, subsequently discovered by Le Verrier himself. They are now much less accurate than the newer tables published by Le Verrier ten years later.

Of Venus new tables were constructed by Mr. Hill in 1872. They are more accurate than any others, being founded on later data than those of Le Verrier, and are therefore satisfactory so far as accuracy of prediction is concerned.

The place of Mars, Jupiter, and Saturn are still computed from the old tables, with certain necessary corrections to make them better represent observations.

The places of Uranus and Neptune are derived from new tables which will probably be sufficiently accurate for some time to come.

For the moon, Pierce's tables have been employed up to the year 1882 inclusive. Commencing with the ephemeris for the year 1883, Hansen's tables are introduced with corrections to the mean longitude founded on two centuries of observation.

With so great a lack of uniformity, and in the absence of any existing tables which have any other element of unity than that of being the work of the same authors, it is extremely desirable that we should be able to compute astronomical ephemerides from a single uniform and consistent set of astronomical data. I hope, in the course of years, to render this possible.

When our ephemeris was first commenced, the corrections applied to existing tables rendered it more accurate than any other. Since that time, the introduction into foreign ephemerides of the improved tables of Le Verrier have rendered them, on the whole, rather more accurate than our own. In one direction, however, our ephemeris will hereafter be far ahead of all others. I mean in its positions of the fixed stars. This portion of it is of particular importance to us, owing to the extent to which our government is engaged in the determination of positions on this continent, and especially in our western territories. Although the places of the stars are determined far more easily than those of the planets, the discussion of star positions has been in almost as backward a state as planetary positions. The errors of old observers have crept in and been continued through two generations of astronomers. A systematic attempt has been made to correct the places of the stars for all systematic errors of this kind, and the work of preparing a catalogue of stars which shall be completely adapted to the determination of time and longitude, both in the fixed observatory and in the field, is now approaching completion. The catalogue cannot be sufficiently complete to give places of the stars for determining the latitude by the zenith telescope, because for such a purpose a much greater number of stars is necessary than can be incorporated in the ephemeris.

From what I have said, it will be seen that the astronomical tables, in general, do not satisfy the scientific condition of completely representing observations to the last degree of accuracy. Few, I think, have an idea how unsystematically work of this kind has hitherto been performed. Until very lately the tables we have possessed have been the work of one man here, another there, and another one somewhere else, each using different methods and different data. The result of this is that there is nothing uniform and systematic among them, and that they have every range of precision. This is no doubt due in part to the fact that the construction of such tables, founded on the mass of observation hitherto made, is entirely beyond the power of any one man. What is wanted is a number of men of different degrees of capacity, all co-operating on a uniform system, so as to obtain a uniform result, like the astronomers in a large observatory. The Greenwich Observatory presents an example of co-operative work of this class extending over more than a century. But it has never extended its operations far outside the field of observation, reduction, and comparison with existing tables. It shows clearly, from time to time, the errors of the tables used in the British Nautical Almanac, but does nothing further, occasional investigations excepted, in the way of supplying new tables. An exception to this is a great work on the theory of the moon's motion, in which Professor Airy is now engaged.

It will be understood that several distinct conditions not yet fulfilled are desirable in astronomical tables; one is that each set of tables shall be founded on absolutely consistent data, for instance, that the masses of the planets shall be the same throughout. Another requirement is that this data shall be as near the truth as astronomical data will suffice to determine them. The third is that the results shall be correct in theory. That is, whether they agree or disagree with observations, they shall be such as result mathematically from the adopted data.

Tables completely fulfilling these conditions are still a work of the future. It is yet to be seen whether such co-operation as is necessary to their production can be secured under any arrangement whatever.

THE WORLD'S DEBT TO ASTRONOMY

Astronomy is more intimately connected than any other science with the history of mankind. While chemistry, physics, and we might say all sciences which pertain to things on the earth, are comparatively modern, we find that contemplative men engaged in the study of the celestial motions even before the commencement of authentic history. The earliest navigators of whom we know must have been aware that the earth was round. This fact was certainly understood by the ancient Greeks and Egyptians, as well as it is at the present day. True, they did not know that the earth revolved on its axis, but thought that the heavens and all that in them is performed a daily revolution around our globe, which was, therefore, the centre of the universe. It was the cynosure, or constellation of the Little Bear, by which the sailors used to guide their ships before the discovery of the mariner's compass. Thus we see both a practical and contemplative side to astronomy through all history. The world owes two debts to that science: one for its practical uses, and the other for the ideas it has afforded us of the immensity of creation.

The practical uses of astronomy are of two kinds: One relates to geography; the other to times, seasons, and chronology. Every navigator who sails long out of sight of land must be something of an astronomer. His compass tells him where are east, west, north, and south, but it gives him no information as to where on the wide ocean he may be, or whither the currents may be carrying him. Even with the swiftest modern steamers it is not safe to trust to the compass in crossing the Atlantic. A number of years ago the steamer City of Washington set out on her usual voyage from Liverpool to New York. By rare bad luck the weather was stormy or cloudy during her whole passage, so that the captain could not get a sight on the sun, and therefore had to trust to his compass and his log-line, the former telling him in what direction he had steamed, and the latter how fast he was going each hour. The result was that the ship ran ashore on the coast of Nova Scotia, when the captain thought he was approaching Nantucket.

Not only the navigator but the surveyor in the western wilds must depend on astronomical observations to learn his exact position on the earth's surface, or the latitude and longitude of the camp which he occupies. He is able to do this because the earth is round, and the direction of the plumb-line not exactly the same at any two places. Let us suppose that the earth stood still, so as not to revolve on its axis at all. Then we should always see the stars at rest and the star which was in the zenith of any place, say a farm-house in New York, at any time, would be there every night and every hour of the year. Now the zenith is simply the point from which the plumb-line seems to drop. Lie on the ground; hang a plummet above your head, sight on the line with one eye, and the direction of the sight will be the zenith of your place. Suppose the earth was still, and a certain star was at your zenith. Then if you went to another place a mile away, the direction of the plumb-line would be slightly different. The change would, indeed, be very small, so small that you could not detect it by sighting with the plumb-line. But astronomers and surveyors have vastly more accurate instruments than the plumb-line and the eye, instruments by which a deviation that the unaided eye could not detect can be seen and measured. Instead of the plumb-line they use a spirit-level or a basin of quicksilver. The surface of quicksilver is exactly level and so at right angles to the true direction of the plumb-line or the force of gravity. Its direction is therefore a little different at two different places on the surface, and the change can be measured by its effect on the apparent direction of a star seen by reflection from the surface.

It is true that a considerable distance on the earth's surface will seem very small in its effect on the position of a star. Suppose there were two stars in the heavens, the one in the zenith of the place where you now stand, and the other in the zenith of a place a mile away. To the best eye unaided by a telescope those two stars would look like a single one. But let the two places be five miles apart, and the eye could see that there were two of them. A good telescope could distinguish between two stars corresponding to places not more than a hundred feet apart. The most exact measurements can determine distances ranging from thirty to sixty feet. If a skilful astronomical observer should mount a telescope on your premises, and determine his latitude by observations on two or three evenings, and then you should try to trick him by taking up the instrument and putting it at another point one hundred feet north or south, he would find out that something was wrong by a single night's work.

Within the past three years a wobbling of the earth's axis has been discovered, which takes place within a circle thirty feet in radius and sixty feet in diameter. Its effect was noticed in astronomical observations many years ago, but the change it produced was so small that men could not find out what the matter was. The exact nature and amount of the wobbling is a work of the exact astronomy of the present time.

We cannot measure across oceans from island to island. Until a recent time we have not even measured across the continent, from New York to San Francisco, in the most precise way. Without astronomy we should know nothing of the distance between New York and Liverpool, except by the time which it took steamers to run it, a measure which would be very uncertain indeed. But by the aid of astronomical observations and the Atlantic cables the distance is found within a few hundred yards. Without astronomy we could scarcely make an accurate map of the United States, except at enormous labor and expense, and even then we could not be sure of its correctness. But the practical astronomer being able to determine his latitude and longitude within fifty yards, the positions of the principal points in all great cities of the country are known, and can be laid down on maps.

The world has always had to depend on astronomy for all its knowledge concerning times and seasons. The changes of the moon gave us the first month, and the year completes its round as the earth travels in its orbit. The results of astronomical observation are for us condensed into almanacs, which are now in such universal use that we never think of their astronomical origin. But in ancient times people had no almanacs, and they learned the time of year, or the number of days in the year, by observing the time when Sirius or some other bright star rose or set with the sun, or disappeared from view in the sun's rays. At Alexandria, in Egypt, the length of the year was determined yet more exactly by observing when the sun rose exactly in the east and set exactly in the west, a date which fixed the equinox for them as for us. More than seventeen hundred years ago, Ptolemy, the great author of The Almagest, had fixed the length of the year to within a very few minutes. He knew it was a little less than 365 1/2 days. The dates of events in ancient history depend very largely on the chronological cycles of astronomy. Eclipses of the sun and moon sometimes fixed the date of great events, and we learn the relation of ancient calendars to our own through the motions of the earth and moon, and can thus measure out the years for the events in ancient history on the same scale that we measure out our own.

At the present day, the work of the practical astronomer is made use of in our daily life throughout the whole country in yet another way. Our fore-fathers had to regulate their clocks by a sundial, or perhaps by a mark at the corner of the house, which showed where the shadow of the house fell at noon. Very rude indeed was this method; and it was uncertain for another reason. It is not always exactly twenty-four hours between two noons by the sun, Sometimes for two or three months the sun will make it noon earlier and earlier every day; and during several other months later and later every day. The result is that, if a clock is perfectly regulated, the sun will be sometimes a quarter of an hour behind it, and sometimes nearly the same amount before it. Any effort to keep the clock in accord with this changing sun was in vain, and so the time of day was always uncertain.

Now, however, at some of the principal observatories of the country astronomical observations are made on every clear night for the express purpose of regulating an astronomical clock with the greatest exactness. Every day at noon a signal is sent to various parts of the country by telegraph, so that all operators and railway men who hear that signal can set their clock at noon within two or three seconds. People who live near railway stations can thus get their time from it, and so exact time is diffused into every household of the land which is at all near a railway station, without the trouble of watching the sun. Thus increased exactness is given to the time on all our railroads, increased safety is obtained, and great loss of time saved to every one. If we estimated the money value of this saving alone we should no doubt find it to be greater than all that our study of astronomy costs.

It must therefore be conceded that, on the whole, astronomy is a science of more practical use than one would at first suppose. To the thoughtless man, the stars seem to have very little relation to his daily life; they might be forever hid from view without his being the worse for it. He wonders what object men can have in devoting themselves to the study of the motions or phenomena of the heavens. But the more he looks into the subject, and the wider the range which his studies include, the more he will be impressed with the great practical usefulness of the science of the heavens. And yet I think it would be a serious error to say that the world's greatest debt to astronomy was owing to its usefulness in surveying, navigation, and chronology. The more enlightened a man is, the more he will feel that what makes his mind what it is, and gives him the ideas of himself and creation which he possesses, is more important than that which gains him wealth. I therefore hold that the world's greatest debt to astronomy is that it has taught us what a great thing creation is, and what an insignificant part of the Creator's work is this earth on which we dwell, and everything that is upon it. That space is infinite, that wherever we go there is a farther still beyond it, must have been accepted as a fact by all men who have thought of the subject since men began to think at all. But it is very curious how hard even the astronomers found it to believe that creation is as large as we now know it to be. The Greeks had their gods on or not very far above Olympus, which was a sort of footstool to the heavens. Sometimes they tried to guess how far it probably was from the vault of heaven to the earth, and they had a myth as to the time it took Vulcan to fall. Ptolemy knew that the moon was about thirty diameters of the earth distant from us, and he knew that the sun was many times farther than the moon; he thought it about twenty times as far, but could not be sure. We know that it is nearly four hundred times as far.

When Copernicus propounded the theory that the earth moved around the sun, and not the sun around the earth, he was able to fix the relative distances of the several planets, and thus make a map of the solar system. But he knew nothing about the scale of this map. He knew, for example, that Venus was a little more than two-thirds the distance of the earth from the sun, and that Mars was about half as far again as the earth, Jupiter about five times, and Saturn about ten times; but he knew nothing about the distance of any one of them from the sun. He had his map all right, but he could not give any scale of miles or any other measurements upon it. The astronomers who first succeeded him found that the distance was very much greater than had formerly been supposed; that it was, in fact, for them immeasurably great, and that was all they could say about it.

The proofs which Copernicus gave that the earth revolved around the sun were so strong that none could well doubt them. And yet there was a difficulty in accepting the theory which seemed insuperable. If the earth really moved in so immense an orbit as it must, then the stars would seem to move in the opposite direction, just as, if you were in a train that is shunting off cars one after another, as the train moves back and forth you see its motion in the opposite motion of every object around you. If then the earth at one side of its orbit was exactly between two stars, when it moved to the other side of its orbit it would not be in a line between them, but each star would have seemed to move in the opposite direction.

For centuries astronomers made the most exact observations that they were able without having succeeded in detecting any such apparent motion among the stars. Here was a mystery which they could not solve. Either the Copernican system was not true, after all, and the earth did not move in an orbit, or the stars were at such immense distances that the whole immeasurable orbit of the earth is a mere point in comparison. Philosophers could not believe that the Creator would waste room by allowing the inconceivable spaces which appeared to lie between our system and the fixed stars to remain unused, and so thought there must be something wrong in the theory of the earth's motion.

Not until the nineteenth century was well in progress did the most skilful observers of their time, Bessel and Struve, having at command the most refined instruments which science was then able to devise, discover the reality of the parallax of the stars, and show that the nearest of these bodies which they could find was more than 400,000 times as far as the 93,000,000 of miles which separate the earth from the sun. During the half-century and more which has elapsed since this discovery, astronomers have been busily engaged in fathoming the heavenly depths. The nearest star they have been able to find is about 280,000 times the sun's distance. A dozen or a score more are within 1,000,000 times that distance. Beyond this all is unfathomable by any sounding-line yet known to man.

The results of these astronomical measures are stupendous beyond conception. No mere statement in numbers conveys any idea of it. Nearly all the brighter stars are known to be flying through space at speeds which generally range between ten and forty or fifty miles per second, some slower and some swifter, even up to one or two hundred miles a second. Such a speed would carry us across the Atlantic while we were reading two or three of these sentences. These motions take place some in one direction and some in another. Some of the stars are coming almost straight towards us. Should they reach us, and pass through our solar system, the result would be destructive to our earth, and perhaps to our sun.

Are we in any danger? No, because, however madly they may come, whether ten, twenty, or one hundred miles per second, so many millions of years must elapse before they reach us that we need give ourselves no concern in the matter. Probably none of them are coming straight to us; their course deviates just a hair's-breadth from our system, but that hair's-breadth is so large a quantity that when the millions of years elapse their course will lie on one side or the other of our system and they will do no harm to our planet; just as a bullet fired at an insect a mile away would be nearly sure to miss it in one direction or the other.

Our instrument makers have constructed telescopes more and more powerful, and with these the whole number of stars visible is carried up into the millions, say perhaps to fifty or one hundred millions. For aught we know every one of those stars may have planets like our own circling round it, and these planets may be inhabited by beings equal to ourselves. To suppose that our globe is the only one thus inhabited is something so unlikely that no one could expect it. It would be very nice to know something about the people who may inhabit these bodies, but we must await our translation to another sphere before we can know anything on the subject. Meanwhile, we have gained what is of more value than gold or silver; we have learned that creation transcends all our conceptions, and our ideas of its Author are enlarged accordingly.

AN ASTRONOMICAL FRIENDSHIP

There are few men with whom I would like so well to have a quiet talk as with Father Hell. I have known more important and more interesting men, but none whose acquaintance has afforded me a serener satisfaction, or imbued me with an ampler measure of a feeling that I am candid enough to call self-complacency. The ties that bind us are peculiar. When I call him my friend, I do not mean that we ever hobnobbed together. But if we are in sympathy, what matters it that he was dead long before I was born, that he lived in one century and I in another? Such differences of generation count for little in the brotherhood of astronomy, the work of whose members so extends through all time that one might well forget that he belongs to one century or to another.

Father Hell was an astronomer. Ask not whether he was a very great one, for in our science we have no infallible gauge by which we try men and measure their stature. He was a lover of science and an indefatigable worker, and he did what in him lay to advance our knowledge of the stars. Let that suffice. I love to fancy that in some other sphere, either within this universe of ours or outside of it, all who have successfully done this may some time gather and exchange greetings. Should this come about there will be a few—Hipparchus and Ptolemy, Copernicus and Newton, Galileo and Herschel—to be surrounded by admiring crowds. But these men will have as warm a grasp and as kind a word for the humblest of their followers, who has merely discovered a comet or catalogued a nebula, as for the more brilliant of their brethren.

My friend wrote the letters S. J. after his name. This would indicate that he had views and tastes which, in some points, were very different from my own. But such differences mark no dividing line in the brotherhood of astronomy. My testimony would count for nothing were I called as witness for the prosecution in a case against the order to which my friend belonged. The record would be very short: Deponent saith that he has at various times known sundry members of the said order; and that they were lovers of sound learning, devoted to the discovery and propagation of knowledge; and further deponent saith not.

If it be true that an undevout astronomer is mad, then was Father Hell the sanest of men. In his diary we find entries like these: "Benedicente Deo, I observed the Sun on the meridian to-day.... Deo quoque benedicente, I to-day got corresponding altitudes of the Sun's upper limb." How he maintained the simplicity of his faith in the true spirit of the modern investigator is shown by his proceedings during a momentous voyage along the coast of Norway, of which I shall presently speak. He and his party were passengers on a Norwegian vessel. For twelve consecutive days they had been driven about by adverse storms, threatened with shipwreck on stony cliffs, and finally compelled to take refuge in a little bay, with another ship bound in the same direction, there to wait for better weather.

Father Hell was philosopher enough to know that unusual events do not happen without cause. Perhaps he would have undergone a week of storm without its occurring to him to investigate the cause of such a bad spell of weather. But when he found the second week approaching its end and yet no sign of the sun appearing or the wind abating, he was satisfied that something must be wrong. So he went to work in the spirit of the modern physician who, when there is a sudden outbreak of typhoid fever, looks at the wells and examines their water with the microscope to find the microbes that must be lurking somewhere. He looked about, and made careful inquiries to find what wickedness captain and crew had been guilty of to bring such a punishment. Success soon rewarded his efforts. The King of Denmark had issued a regulation that no fish or oil should be sold along the coast except by the regular dealers in those articles. And the vessel had on board contraband fish and blubber, to be disposed of in violation of this law.

The astronomer took immediate and energetic measures to insure the public safety. He called the crew together, admonished them of their sin, the suffering they were bringing on themselves, and the necessity of getting back to their families. He exhorted them to throw the fish overboard, as the only measure to secure their safety. In the goodness of his heart, he even offered to pay the value of the jettison as soon as the vessel reached Drontheim.

But the descendants of the Vikings were stupid and unenlightened men— "educatione sua et professione homines crassissimi"—and would not swallow the medicine so generously offered. They claimed that, as they had bought the fish from the Russians, their proceedings were quite lawful. As for being paid to throw the fish overboard, they must have spot cash in advance or they would not do it.

After further fruitless conferences, Father Hell determined to escape the danger by transferring his party to the other vessel. They had not more than got away from the wicked crew than Heaven began to smile on their act—"factum comprobare Deus ipse videtur"—the clouds cleared away, the storm ceased to rage, and they made their voyage to Copenhagen under sunny skies. I regret to say that the narrative is silent as to the measure of storm subsequently awarded to the homines crassissimi of the forsaken vessel.

For more than a century Father Hell had been a well-known figure in astronomical history. His celebrity was not, however, of such a kind as the Royal Astronomer of Austria that he was ought to enjoy. A not unimportant element in his fame was a suspicion of his being a black sheep in the astronomical flock. He got under this cloud through engaging in a trying and worthy enterprise. On June 3, 1769, an event occurred which had for generations been anticipated with the greatest interest by the whole astronomical world. This was a transit of Venus over the disk of the sun. Our readers doubtless know that at that time such a transit afforded the most accurate method known of determining the distance of the earth from the sun. To attain this object, parties were sent to the most widely separated parts of the globe, not only over wide stretches of longitude, but as near as possible to the two poles of the earth. One of the most favorable and important regions of observation was Lapland, and the King of Denmark, to whom that country then belonged, interested himself in getting a party sent thither. After a careful survey of the field he selected Father Hell, Chief of the Observatory at Vienna, and well known as editor and publisher of an annual ephemeris, in which the movements and aspects of the heavenly bodies were predicted. The astronomer accepted the mission and undertook what was at that time a rather hazardous voyage. His station was at Vardo in the region of the North Cape. What made it most advantageous for the purpose was its being situated several degrees within the Arctic Circle, so that on the date of the transit the sun did not set. The transit began when the sun was still two or three hours from his midnight goal, and it ended nearly an equal time afterwards. The party consisted of Hell himself, his friend and associate, Father Sajnovics, one Dominus Borgrewing, of whom history, so far as I know, says nothing more, and an humble individual who in the record receives no other designation than "Familias." This implies, we may suppose, that he pitched the tent and made the coffee. If he did nothing but this we might pass him over in silence. But we learn that on the day of the transit he stood at the clock and counted the all-important seconds while the observations were going on.

The party was favored by cloudless weather, and made the required observations with entire success. They returned to Copenhagen, and there

Father Hell remained to edit and publish his work. Astronomers were naturally anxious to get the results, and showed some impatience when it became known that Hell refused to announce them until they were all reduced and printed in proper form under the auspices of his royal patron. While waiting, the story got abroad that he was delaying the work until he got the results of observations made elsewhere, in order to "doctor" his own and make them fit in with the others. One went so far as to express a suspicion that Hell had not seen the transit at all, owing to clouds, and that what he pretended to publish were pure fabrications. But his book came out in a few months in such good form that this suspicion was evidently groundless. Still, the fears that the observations were not genuine were not wholly allayed, and the results derived from them were, in consequence, subject to some doubt. Hell himself considered the reflections upon his integrity too contemptible to merit a serious reply. It is said that he wrote to some one offering to exhibit his journal free from interlineations or erasures, but it does not appear that there is any sound authority for this statement. What is of some interest is that he published a determination of the parallax of the sun based on the comparison of his own observations with those made at other stations. The result was 8".70. It was then, and long after, supposed that the actual value of the parallax was about 8".50, and the deviation of Hell's result from this was considered to strengthen the doubt as to the correctness of his work. It is of interest to learn that, by the most recent researches, the number in question must be between 8".75 and 8".80, so that in reality Hell's computations came nearer the truth than those generally current during the century following his work.

Thus the matter stood for sixty years after the transit, and for a generation after Father Hell had gone to his rest. About 1830 it was found that the original journal of his voyage, containing the record of his work as first written down at the station, was still preserved at the Vienna Observatory. Littrow, then an astronomer at Vienna, made a critical examination of this record in order to determine whether it had been tampered with. His conclusions were published in a little book giving a transcript of the journal, a facsimile of the most important entries, and a very critical description of the supposed alterations made in them. He reported in substance that the original record had been so tampered with that it was impossible to decide whether the observations as published were genuine or not. The vital figures, those which told the times when Venus entered upon the sun, had been erased, and rewritten with blacker ink. This might well have been done after the party returned to Copenhagen. The case against the observer seemed so well made out that professors of astronomy gave their hearers a lesson in the value of truthfulness, by telling them how Father Hell had destroyed what might have been very good observations by trying to make them appear better than they really were.

In 1883 I paid a visit to Vienna for the purpose of examining the great telescope which had just been mounted in the observatory there by Grubb, of Dublin. The weather was so unfavorable that it was necessary to remain two weeks, waiting for an opportunity to see the stars. One evening I visited the theatre to see Edwin Booth, in his celebrated tour over the Continent, play King Lear to the applauding Viennese. But evening amusements cannot be utilized to kill time during the day. Among the works I had projected was that of rediscussing all the observations made on the transits of Venus which had occurred in 1761 and 1769, by the light of modern discovery. As I have already remarked, Hell's observations were among the most important made, if they were only genuine. So, during my almost daily visits to the observatory, I asked permission of the director to study Hell's manuscript, which was deposited in the library of the institution. Permission was freely given, and for some days I pored over the manuscript. It is a very common experience in scientific research that a subject which seems very unpromising when first examined may be found more and more interesting as one looks further into it. Such was the case here. For some time there did not seem any possibility of deciding the question whether the record was genuine. But every time I looked at it some new point came to light. I compared the pages with Littrow's published description and was struck by a seeming want of precision, especially when he spoke of the ink with which the record had been made. Erasers were doubtless unknown in those days—at least our astronomer had none on his expedition—so when he found he had written the wrong word he simply wiped the place off with, perhaps, his finger and wrote what he wanted to say. In such a case Littrow described the matter as erased and new matter written. When the ink flowed freely from the quill pen it was a little dark. Then Littrow said a different kind of ink had been used, probably after he had got back from his journey. On the other hand, there was a very singular case in which there had been a subsequent interlineation in ink of quite a different tint, which Littrow said nothing about. This seemed so curious that I wrote in my notes as follows:

"That Littrow, in arraying his proofs of Hell's forgery, should have failed to dwell upon the obvious difference between this ink and that with which the alterations were made leads me to suspect a defect in his sense of color."

The more I studied the description and the manuscript the stronger this impression became. Then it occurred to me to inquire whether perhaps such could have been the case. So I asked Director Weiss whether anything was known as to the normal character of Littrow's power of distinguishing colors. His answer was prompt and decisive. "Oh yes, Littrow was color-blind to red. He could not distinguish between the color of Aldebaran and the whitest star." No further research was necessary. For half a century the astronomical world had based an impression on the innocent but mistaken evidence of a color-blind man—respecting the tints of ink in a manuscript. It has doubtless happened more than once that when an intimate friend has suddenly and unexpectedly passed away, the reader has ardently wished that it were possible to whisper just one word of appreciation across the dark abyss. And so it is that I have ever since felt that I would like greatly to tell Father Hell the story of my work at Vienna in 1883.

THE EVOLUTION OF THE SCIENTIFIC INVESTIGATOR

[Footnote: Presidential address at the opening of the International Congress of Arts and Science, St. Louis Exposition, September 21: 1904.]

As we look at the assemblage gathered in this hall, comprising so many names of widest renown in every branch of learning—we might almost say in every field of human endeavor—the first inquiry suggested must be after the object of our meeting. The answer is that our purpose corresponds to the eminence of the assemblage. We aim at nothing less than a survey of the realm of knowledge, as comprehensive as is permitted by the limitations of time and space. The organizers of our congress have honored me with the charge of presenting such preliminary view of its field as may make clear the spirit of our undertaking.

Certain tendencies characteristic of the science of our day clearly suggest the direction of our thoughts most appropriate to the occasion. Among the strongest of these is one towards laying greater stress on questions of the beginnings of things, and regarding a knowledge of the laws of development of any object of study as necessary to the understanding of its present form. It may be conceded that the principle here involved is as applicable in the broad field before us as in a special research into the properties of the minutest organism. It therefore seems meet that we should begin by inquiring what agency has brought about the remarkable development of science to which the world of to-day bears witness. This view is recognized in the plan of our proceedings by providing for each great department of knowledge a review of its progress during the century that has elapsed since the great event commemorated by the scenes outside this hall. But such reviews do not make up that general survey of science at large which is necessary to the development of our theme, and which must include the action of causes that had their origin long before our time. The movement which culminated in making the nineteenth century ever memorable in history is the outcome of a long series of causes, acting through many centuries, which are worthy of especial attention on such an occasion as this. In setting them forth we should avoid laying stress on those visible manifestations which, striking the eye of every beholder, are in no danger of being overlooked, and search rather for those agencies whose activities underlie the whole visible scene, but which are liable to be blotted out of sight by the very brilliancy of the results to which they have given rise. It is easy to draw attention to the wonderful qualities of the oak; but, from that very fact, it may be needful to point out that the real wonder lies concealed in the acorn from which it grew.

Our inquiry into the logical order of the causes which have made our civilization what it is to-day will be facilitated by bringing to mind certain elementary considerations—ideas so familiar that setting them forth may seem like citing a body of truisms—and yet so frequently overlooked, not only individually, but in their relation to each other, that the conclusion to which they lead may be lost to sight. One of these propositions is that psychical rather than material causes are those which we should regard as fundamental in directing the development of the social organism. The human intellect is the really active agent in every branch of endeavor—the primum mobile of civilization—and all those material manifestations to which our attention is so often directed are to be regarded as secondary to this first agency. If it be true that "in the world is nothing great but man; in man is nothing great but mind," then should the key-note of our discourse be the recognition of this first and greatest of powers.

Another well-known fact is that those applications of the forces of nature to the promotion of human welfare which have made our age what it is are of such comparatively recent origin that we need go back only a single century to antedate their most important features, and scarcely more than four centuries to find their beginning. It follows that the subject of our inquiry should be the commencement, not many centuries ago, of a certain new form of intellectual activity.

Having gained this point of view, our next inquiry will be into the nature of that activity and its relation to the stages of progress which preceded and followed its beginning. The superficial observer, who sees the oak but forgets the acorn, might tell us that the special qualities which have brought out such great results are expert scientific knowledge and rare ingenuity, directed to the application of the powers of steam and electricity. From this point of view the great inventors and the great captains of industry were the first agents in bringing about the modern era. But the more careful inquirer will see that the work of these men was possible only through a knowledge of the laws of nature, which had been gained by men whose work took precedence of theirs in logical order, and that success in invention has been measured by completeness in such knowledge. While giving all due honor to the great inventors, let us remember that the first place is that of the great investigators, whose forceful intellects opened the way to secrets previously hidden from men. Let it be an honor and not a reproach to these men that they were not actuated by the love of gain, and did not keep utilitarian ends in view in the pursuit of their researches. If it seems that in neglecting such ends they were leaving undone the most important part of their work, let us remember that Nature turns a forbidding face to those who pay her court with the hope of gain, and is responsive only to those suitors whose love for her is pure and undefiled. Not only is the special genius required in the investigator not that generally best adapted to applying the discoveries which he makes, but the result of his having sordid ends in view would be to narrow the field of his efforts, and exercise a depressing effect upon his activities. The true man of science has no such expression in his vocabulary as "useful knowledge." His domain is as wide as nature itself, and he best fulfils his mission when he leaves to others the task of applying the knowledge he gives to the world.

We have here the explanation of the well-known fact that the functions of the investigator of the laws of nature, and of the inventor who applies these laws to utilitarian purposes, are rarely united in the same person. If the one conspicuous exception which the past century presents to this rule is not unique, we should probably have to go back to Watt to find another.

From this view-point it is clear that the primary agent in the movement which has elevated man to the masterful position he now occupies is the scientific investigator. He it is whose work has deprived plague and pestilence of their terrors, alleviated human suffering, girdled the earth with the electric wire, bound the continent with the iron way, and made neighbors of the most distant nations. As the first agent which has made possible this meeting of his representatives, let his evolution be this day our worthy theme. As we follow the evolution of an organism by studying the stages of its growth, so we have to show how the work of the scientific investigator is related to the ineffectual efforts of his predecessors.

In our time we think of the process of development in nature as one going continuously forward through the combination of the opposite processes of evolution and dissolution. The tendency of our thought has been in the direction of banishing cataclysms to the theological limbo, and viewing Nature as a sleepless plodder, endowed with infinite patience, waiting through long ages for results. I do not contest the truth of the principle of continuity on which this view is based. But it fails to make known to us the whole truth. The building of a ship from the time that her keel is laid until she is making her way across the ocean is a slow and gradual process; yet there is a cataclysmic epoch opening up a new era in her history. It is the moment when, after lying for months or years a dead, inert, immovable mass, she is suddenly endowed with the power of motion, and, as if imbued with life, glides into the stream, eager to begin the career for which she was designed.

I think it is thus in the development of humanity. Long ages may pass during which a race, to all external observation, appears to be making no real progress. Additions may be made to learning, and the records of history may constantly grow, but there is nothing in its sphere of thought, or in the features of its life, that can be called essentially new. Yet, Nature may have been all along slowly working in a way which evades our scrutiny, until the result of her operations suddenly appears in a new and revolutionary movement, carrying the race to a higher plane of civilization.

It is not difficult to point out such epochs in human progress. The greatest of all, because it was the first, is one of which we find no record either in written or geological history. It was the epoch when our progenitors first took conscious thought of the morrow, first used the crude weapons which Nature had placed within their reach to kill their prey, first built a fire to warm their bodies and cook their food. I love to fancy that there was some one first man, the Adam of evolution, who did all this, and who used the power thus acquired to show his fellows how they might profit by his example. When the members of the tribe or community which he gathered around him began to conceive of life as a whole—to include yesterday, to-day, and to-morrow in the same mental grasp—to think how they might apply the gifts of Nature to their own uses—a movement was begun which should ultimately lead to civilization.

Long indeed must have been the ages required for the development of this rudest primitive community into the civilization revealed to us by the most ancient tablets of Egypt and Assyria. After spoken language was developed, and after the rude representation of ideas by visible marks drawn to resemble them had long been practised, some Cadmus must have invented an alphabet. When the use of written language was thus introduced, the word of command ceased to be confined to the range of the human voice, and it became possible for master minds to extend their influence as far as a written message could be carried. Then were communities gathered into provinces; provinces into kingdoms, kingdoms into great empires of antiquity. Then arose a stage of civilization which we find pictured in the most ancient records—a stage in which men were governed by laws that were perhaps as wisely adapted to their conditions as our laws are to ours— in which the phenomena of nature were rudely observed, and striking occurrences in the earth or in the heavens recorded in the annals of the nation.

Vast was the progress of knowledge during the interval between these empires and the century in which modern science began. Yet, if I am right in making a distinction between the slow and regular steps of progress, each growing naturally out of that which preceded it, and the entrance of the mind at some fairly definite epoch into an entirely new sphere of activity, it would appear that there was only one such epoch during the entire interval. This was when abstract geometrical reasoning commenced, and astronomical observations aiming at precision were recorded, compared, and discussed. Closely associated with it must have been the construction of the forms of logic. The radical difference between the demonstration of a theorem of geometry and the reasoning of every-day life which the masses of men must have practised from the beginning, and which few even to-day ever get beyond, is so evident at a glance that I need not dwell upon it. The principal feature of this advance is that, by one of those antinomies of human intellect of which examples are not wanting even in our own time, the development of abstract ideas preceded the concrete knowledge of natural phenomena. When we reflect that in the geometry of Euclid the science of space was brought to such logical perfection that even to-day its teachers are not agreed as to the practicability of any great improvement upon it, we cannot avoid the feeling that a very slight change in the direction of the intellectual activity of the Greeks would have led to the beginning of natural science. But it would seem that the very purity and perfection which was aimed at in their system of geometry stood in the way of any extension or application of its methods and spirit to the field of nature. One example of this is worthy of attention. In modern teaching the idea of magnitude as generated by motion is freely introduced. A line is described by a moving point; a plane by a moving line; a solid by a moving plane. It may, at first sight, seem singular that this conception finds no place in the Euclidian system. But we may regard the omission as a mark of logical purity and rigor. Had the real or supposed advantages of introducing motion into geometrical conceptions been suggested to Euclid, we may suppose him to have replied that the theorems of space are independent of time; that the idea of motion necessarily implies time, and that, in consequence, to avail ourselves of it would be to introduce an extraneous element into geometry.

It is quite possible that the contempt of the ancient philosophers for the practical application of their science, which has continued in some form to our own time, and which is not altogether unwholesome, was a powerful factor in the same direction. The result was that, in keeping geometry pure from ideas which did not belong to it, it failed to form what might otherwise have been the basis of physical science. Its founders missed the discovery that methods similar to those of geometric demonstration could be extended into other and wider fields than that of space. Thus not only the development of applied geometry but the reduction of other conceptions to a rigorous mathematical form was indefinitely postponed.

There is, however, one science which admitted of the immediate application of the theorems of geometry, and which did not require the application of the experimental method. Astronomy is necessarily a science of observation pure and simple, in which experiment can have no place except as an auxiliary. The vague accounts of striking celestial phenomena handed down by the priests and astrologers of antiquity were followed in the time of the Greeks by observations having, in form at least, a rude approach to precision, though nothing like the degree of precision that the astronomer of to-day would reach with the naked eye, aided by such instruments as he could fashion from the tools at the command of the ancients.

The rude observations commenced by the Babylonians were continued with gradually improving instruments—first by the Greeks and afterwards by the Arabs—but the results failed to afford any insight into the true relation of the earth to the heavens. What was most remarkable in this failure is that, to take a first step forward which would have led on to success, no more was necessary than a course of abstract thinking vastly easier than that required for working out the problems of geometry. That space is infinite is an unexpressed axiom, tacitly assumed by Euclid and his successors. Combining this with the most elementary consideration of the properties of the triangle, it would be seen that a body of any given size could be placed at such a distance in space as to appear to us like a point. Hence a body as large as our earth, which was known to be a globe from the time that the ancient Phoenicians navigated the Mediterranean, if placed in the heavens at a sufficient distance, would look like a star. The obvious conclusion that the stars might be bodies like our globe, shining either by their own light or by that of the sun, would have been a first step to the understanding of the true system of the world.

There is historic evidence that this deduction did not wholly escape the Greek thinkers. It is true that the critical student will assign little weight to the current belief that the vague theory of Pythagoras—that fire was at the centre of all things—implies a conception of the heliocentric theory of the solar system. But the testimony of Archimedes, confused though it is in form, leaves no serious doubt that Aristarchus of Samos not only propounded the view that the earth revolves both on its own axis and around the sun, but that he correctly removed the great stumbling-block in the way of this theory by adding that the distance of the fixed stars was infinitely greater than the dimensions of the earth's orbit. Even the world of philosophy was not yet ready for this conception, and, so far from seeing the reasonableness of the explanation, we find Ptolemy arguing against the rotation of the earth on grounds which careful observations of the phenomena around him would have shown to be ill-founded.

Physical science, if we can apply that term to an uncoordinated body of facts, was successfully cultivated from the earliest times. Something must have been known of the properties of metals, and the art of extracting them from their ores must have been practised, from the time that coins and medals were first stamped. The properties of the most common compounds were discovered by alchemists in their vain search for the philosopher's stone, but no actual progress worthy of the name rewarded the practitioners of the black art.

Perhaps the first approach to a correct method was that of Archimedes, who by much thinking worked out the law of the lever, reached the conception of the centre of gravity, and demonstrated the first principles of hydrostatics. It is remarkable that he did not extend his researches into the phenomena of motion, whether spontaneous or produced by force. The stationary condition of the human intellect is most strikingly illustrated by the fact that not until the time of Leonardo was any substantial advance made on his discovery. To sum up in one sentence the most characteristic feature of ancient and medieval science, we see a notable contrast between the precision of thought implied in the construction and demonstration of geometrical theorems and the vague indefinite character of the ideas of natural phenomena generally, a contrast which did not disappear until the foundations of modern science began to be laid.

We should miss the most essential point of the difference between medieval and modern learning if we looked upon it as mainly a difference either in the precision or the amount of knowledge. The development of both of these qualities would, under any circumstances, have been slow and gradual, but sure. We can hardly suppose that any one generation, or even any one century, would have seen the complete substitution of exact for inexact ideas. Slowness of growth is as inevitable in the case of knowledge as in that of a growing organism. The most essential point of difference is one of those seemingly slight ones, the importance of which we are too apt to overlook. It was like the drop of blood in the wrong place, which some one has told us makes all the difference between a philosopher and a maniac. It was all the difference between a living tree and a dead one, between an inert mass and a growing organism. The transition of knowledge from the dead to the living form must, in any complete review of the subject, be looked upon as the really great event of modern times. Before this event the intellect was bound down by a scholasticism which regarded knowledge as a rounded whole, the parts of which were written in books and carried in the minds of learned men. The student was taught from the beginning of his work to look upon authority as the foundation of his beliefs. The older the authority the greater the weight it carried. So effective was this teaching that it seems never to have occurred to individual men that they had all the opportunities ever enjoyed by Aristotle of discovering truth, with the added advantage of all his knowledge to begin with. Advanced as was the development of formal logic, that practical logic was wanting which could see that the last of a series of authorities, every one of which rested on those which preceded it, could never form a surer foundation for any doctrine than that supplied by its original propounder.

The result of this view of knowledge was that, although during the fifteen centuries following the death of the geometer of Syracuse great universities were founded at which generations of professors expounded all the learning of their time, neither professor nor student ever suspected what latent possibilities of good were concealed in the most familiar operations of Nature. Every one felt the wind blow, saw water boil, and heard the thunder crash, but never thought of investigating the forces here at play. Up to the middle of the fifteenth century the most acute observer could scarcely have seen the dawn of a new era.

In view of this state of things it must be regarded as one of the most remarkable facts in evolutionary history that four or five men, whose mental constitution was either typical of the new order of things, or who were powerful agents in bringing it about, were all born during the fifteenth century, four of them at least, at so nearly the same time as to be contemporaries.

Leonardo da Vinci, whose artistic genius has charmed succeeding generations, was also the first practical engineer of his time, and the first man after Archimedes to make a substantial advance in developing the laws of motion. That the world was not prepared to make use of his scientific discoveries does not detract from the significance which must attach to the period of his birth.

Shortly after him was born the great navigator whose bold spirit was to make known a new world, thus giving to commercial enterprise that impetus which was so powerful an agent in bringing about a revolution in the thoughts of men.

The birth of Columbus was soon followed by that of Copernicus, the first after Aristarchus to demonstrate the true system of the world. In him more than in any of his contemporaries do we see the struggle between the old forms of thought and the new. It seems almost pathetic and is certainly most suggestive of the general view of knowledge taken at that time that, instead of claiming credit for bringing to light great truths before unknown, he made a labored attempt to show that, after all, there was nothing really new in his system, which he claimed to date from Pythagoras and Philolaus. In this connection it is curious that he makes no mention of Aristarchus, who I think will be regarded by conservative historians as his only demonstrated predecessor. To the hold of the older ideas upon his mind we must attribute the fact that in constructing his system he took great pains to make as little change as possible in ancient conceptions.

Luther, the greatest thought-stirrer of them all, practically of the same generation with Copernicus, Leonardo and Columbus, does not come in as a scientific investigator, but as the great loosener of chains which had so fettered the intellect of men that they dared not think otherwise than as the authorities thought.

Almost coeval with the advent of these intellects was the invention of printing with movable type. Gutenberg was born during the first decade of the century, and his associates and others credited with the invention not many years afterwards. If we accept the principle on which I am basing my argument, that in bringing out the springs of our progress we should assign the first place to the birth of those psychic agencies which started men on new lines of thought, then surely was the fifteenth the wonderful century.

Let us not forget that, in assigning the actors then born to their places, we are not narrating history, but studying a special phase of evolution. It matters not for us that no university invited Leonardo to its halls, and that his science was valued by his contemporaries only as an adjunct to the art of engineering. The great fact still is that he was the first of mankind to propound laws of motion. It is not for anything in Luther's doctrines that he finds a place in our scheme. No matter for us whether they were sound or not. What he did towards the evolution of the scientific investigator was to show by his example that a man might question the best-established and most venerable authority and still live—still preserve his intellectual integrity—still command a hearing from nations and their rulers. It matters not for us whether Columbus ever knew that he had discovered a new continent. His work was to teach that neither hydra, chimera nor abyss—neither divine injunction nor infernal machination—was in the way of men visiting every part of the globe, and that the problem of conquering the world reduced itself to one of sails and rigging, hull and compass. The better part of Copernicus was to direct man to a view-point whence he should see that the heavens were of like matter with the earth. All this done, the acorn was planted from which the oak of our civilization should spring. The mad quest for gold which followed the discovery of Columbus, the questionings which absorbed the attention of the learned, the indignation excited by the seeming vagaries of a Paracelsus, the fear and trembling lest the strange doctrine of Copernicus should undermine the faith of centuries, were all helps to the germination of the seed—stimuli to thought which urged it on to explore the new fields opened up to its occupation. This given, all that has since followed came out in regular order of development, and need be here considered only in those phases having a special relation to the purpose of our present meeting.

So slow was the growth at first that the sixteenth century may scarcely have recognized the inauguration of a new era. Torricelli and Benedetti were of the third generation after Leonardo, and Galileo, the first to make a substantial advance upon his theory, was born more than a century after him. Only two or three men appeared in a generation who, working alone, could make real progress in discovery, and even these could do little in leavening the minds of their fellowmen with the new ideas.

Up to the middle of the seventeenth century an agent which all experience since that time shows to be necessary to the most productive intellectual activity was wanting. This was the attrition of like minds, making suggestions to one another, criticising, comparing, and reasoning. This element was introduced by the organization of the Royal Society of London and the Academy of Sciences of Paris.

The members of these two bodies seem like ingenious youth suddenly thrown into a new world of interesting objects, the purposes and relations of which they had to discover. The novelty of the situation is strikingly shown in the questions which occupied the minds of the incipient investigators. One natural result of British maritime enterprise was that the aspirations of the Fellows of the Royal Society were not confined to any continent or hemisphere. Inquiries were sent all the way to Batavia to know "whether there be a hill in Sumatra which burneth continually, and a fountain which runneth pure balsam." The astronomical precision with which it seemed possible that physiological operations might go on was evinced by the inquiry whether the Indians can so prepare that stupefying herb Datura that "they make it lie several days, months, years, according as they will, in a man's body without doing him any harm, and at the end kill him without missing an hour's time." Of this continent one of the inquiries was whether there be a tree in Mexico that yields water, wine, vinegar, milk, honey, wax, thread and needles.

Among the problems before the Paris Academy of Sciences those of physiology and biology took a prominent place. The distillation of compounds had long been practised, and the fact that the more spirituous elements of certain substances were thus separated naturally led to the question whether the essential essences of life might not be discoverable in the same way. In order that all might participate in the experiments, they were conducted in open session of the academy, thus guarding against the danger of any one member obtaining for his exclusive personal use a possible elixir of life. A wide range of the animal and vegetable kingdom, including cats, dogs and birds of various species, were thus analyzed. The practice of dissection was introduced on a large scale. That of the cadaver of an elephant occupied several sessions, and was of such interest that the monarch himself was a spectator.

To the same epoch with the formation and first work of these two bodies belongs the invention of a mathematical method which in its importance to the advance of exact science may be classed with the invention of the alphabet in its relation to the progress of society at large. The use of algebraic symbols to represent quantities had its origin before the commencement of the new era, and gradually grew into a highly developed form during the first two centuries of that era. But this method could represent quantities only as fixed. It is true that the elasticity inherent in the use of such symbols permitted of their being applied to any and every quantity; yet, in any one application, the quantity was considered as fixed and definite. But most of the magnitudes of nature are in a state of continual variation; indeed, since all motion is variation, the latter is a universal characteristic of all phenomena. No serious advance could be made in the application of algebraic language to the expression of physical phenomena until it could be so extended as to express variation in quantities, as well as the quantities themselves. This extension, worked out independently by Newton and Leibnitz, may be classed as the most fruitful of conceptions in exact science. With it the way was opened for the unimpeded and continually accelerated progress of the last two centuries.

The feature of this period which has the closest relation to the purpose of our coming together is the seemingly unending subdivision of knowledge into specialties, many of which are becoming so minute and so isolated that they seem to have no interest for any but their few pursuers. Happily science itself has afforded a corrective for its own tendency in this direction. The careful thinker will see that in these seemingly diverging branches common elements and common principles are coming more and more to light. There is an increasing recognition of methods of research, and of deduction, which are common to large branches, or to the whole of science. We are more and more recognizing the principle that progress in knowledge implies its reduction to more exact forms, and the expression of its ideas in language more or less mathematical. The problem before the organizers of this Congress was, therefore, to bring the sciences together, and seek for the unity which we believe underlies their infinite diversity.

The assembling of such a body as now fills this hall was scarcely possible in any preceding generation, and is made possible now only through the agency of science itself. It differs from all preceding international meetings by the universality of its scope, which aims to include the whole of knowledge. It is also unique in that none but leaders have been sought out as members. It is unique in that so many lands have delegated their choicest intellects to carry on its work. They come from the country to which our republic is indebted for a third of its territory, including the ground on which we stand; from the land which has taught us that the most scholarly devotion to the languages and learning of the cloistered past is compatible with leadership in the practical application of modern science to the arts of life; from the island whose language and literature have found a new field and a vigorous growth in this region; from the last seat of the holy Roman Empire; from the country which, remembering a monarch who made an astronomical observation at the Greenwich Observatory, has enthroned science in one of the highest places in its government; from the peninsula so learned that we have invited one of its scholars to come and tells us of our own language; from the land which gave birth to Leonardo, Galileo, Torricelli, Columbus, Volta—what an array of immortal names!—from the little republic of glorious history which, breeding men rugged as its eternal snow-peaks, has yet been the seat of scientific investigation since the day of the Bernoullis; from the land whose heroic dwellers did not hesitate to use the ocean itself to protect it against invaders, and which now makes us marvel at the amount of erudition compressed within its little area; from the nation across the Pacific, which, by half a century of unequalled progress in the arts of life, has made an important contribution to evolutionary science through demonstrating the falsity of the theory that the most ancient races are doomed to be left in the rear of the advancing age—in a word, from every great centre of intellectual activity on the globe I see before me eminent representatives of that world—advance in knowledge which we have met to celebrate. May we not confidently hope that the discussions of such an assemblage will prove pregnant of a future for science which shall outshine even its brilliant past.

Gentlemen and scholars all! You do not visit our shores to find great collections in which centuries of humanity have given expression on canvas and in marble to their hopes, fears, and aspirations. Nor do you expect institutions and buildings hoary with age. But as you feel the vigor latent in the fresh air of these expansive prairies, which has collected the products of human genius by which we are here surrounded, and, I may add, brought us together; as you study the institutions which we have founded for the benefit, not only of our own people, but of humanity at large; as you meet the men who, in the short space of one century, have transformed this valley from a savage wilderness into what it is today—then may you find compensation for the want of a past like yours by seeing with prophetic eye a future world-power of which this region shall be the seat. If such is to be the outcome of the institutions Which we are now building up, then may your present visit be a blessing both to your posterity and ours by making that power one for good to all man-kind. Your deliberations will help to demonstrate to us and to the world at large that the reign of law must supplant that of brute force in the relations of the nations, just as it has supplanted it in the relations of individuals. You will help to show that the war which science is now waging against the sources of diseases, pain, and misery offers an even nobler field for the exercise of heroic qualities than can that of battle. We hope that when, after your all too-fleeting sojourn in our midst, you return to your own shores, you will long feel the influence of the new air you have breathed in an infusion of increased vigor in pursuing your varied labors. And if a new impetus is thus given to the great intellectual movement of the past century, resulting not only in promoting the unification of knowledge, but in widening its field through new combinations of effort on the part of its votaries, the projectors, organizers and supporters of this Congress of Arts and Science will be justified of their labors.

THE EVOLUTION OF ASTRONOMICAL KNOWLEDGE

[Footnote: Address at the dedication of the Flower Observatory, University of Pennsylvania, May 12, 1897—Science, May 21, 1897]
Assembled, as we are, to dedicate a new institution to the promotion of our knowledge of the heavens, it appeared to me that an appropriate and interesting subject might be the present and future problems of astronomy. Yet it seemed, on further reflection, that, apart from the difficulty of making an adequate statement of these problems on such an occasion as the present, such a wording of the theme would not fully express the idea which I wish to convey. The so-called problems of astronomy are not separate and independent, but are rather the parts of one great problem, that of increasing our knowledge of the universe in its widest extent. Nor is it easy to contemplate the edifice of astronomical science as it now stands, without thinking of the past as well as of the present and future. The fact is that our knowledge of the universe has been in the nature of a slow and gradual evolution, commencing at a very early period in human history, and destined to go forward without stop, as we hope, so long as civilization shall endure. The astronomer of every age has built on the foundations laid by his predecessors, and his work has always formed, and must ever form, the base on which his successors shall build. The astronomer of to-day may look back upon Hipparchus and Ptolemy as the earliest ancestors of whom he has positive knowledge. He can trace his scientific descent from generation to generation, through the periods of Arabian and medieval science, through Copernicus, Kepler, Newton, Laplace, and Herschel, down to the present time. The evolution of astronomical knowledge, generally slow and gradual, offering little to excite the attention of the public, has yet

been marked by two cataclysms. One of these is seen in the grand conception of Copernicus that this earth on which we dwell is not a globe fixed in the centre of the universe, but is simply one of a number of bodies, turning on their own axes and at the same time moving around the sun as a centre. It has always seemed to me that the real significance of the heliocentric system lies in the greatness of this conception rather than in the fact of the discovery itself. There is no figure in astronomical history which may more appropriately claim the admiration of mankind through all time than that of Copernicus. Scarcely any great work was ever so exclusively the work of one man as was the heliocentric system the work of the retiring sage of Frauenburg. No more striking contrast between the views of scientific research entertained in his time and in ours can be found than that afforded by the fact that, instead of claiming credit for his great work, he deemed it rather necessary to apologize for it and, so far as possible, to attribute his ideas to the ancients.

A century and a half after Copernicus followed the second great step, that taken by Newton. This was nothing less than showing that the seemingly complicated and inexplicable motions of the heavenly bodies were only special cases of the same kind of motion, governed by the same forces, that we see around us whenever a stone is thrown by the hand or an apple falls to the ground. The actual motions of the heavens and the laws which govern them being known, man had the key with which he might commence to unlock the mysteries of the universe.

When Huyghens, in 1656, published his Systema Saturnium, where he first set forth the mystery of the rings of Saturn, which, for nearly half a century, had perplexed telescopic observers, he prefaced it with a remark that many, even among the learned, might condemn his course in devoting so much time and attention to matters far outside the earth, when he might better be studying subjects of more concern to humanity. Notwithstanding that the inventor of the pendulum clock was, perhaps, the last astronomer against whom a neglect of things terrestrial could be charged, he thought it necessary to enter into an elaborate defence of his course in studying the heavens. Now, however, the more distant objects are in space—I might almost add the more distant events are in time—the more they excite the attention of the astronomer, if only he can hope to acquire positive knowledge about them. Not, however, because he is more interested in things distant than in things near, but because thus he may more completely embrace in the scope of his work the beginning and the end, the boundaries of all things, and thus, indirectly, more fully comprehend all that they include. From his stand-point,

"All are but parts of one stupendous whole,
Whose body Nature is and God the soul."

Others study Nature and her plans as we see them developed on the surface of this little planet which we inhabit, the astronomer would fain learn the plan on which the whole universe is constructed. The magnificent conception of Copernicus is, for him, only an introduction to the yet more magnificent conception of infinite space containing a collection of bodies which we call the visible universe. How far does this universe extend? What are the distances and arrangements of the stars? Does the universe constitute a system? If so, can we comprehend the plan on which this system is formed, of its beginning and of its end? Has it bounds outside of which nothing exists but the black and starless depths of infinity itself? Or are the stars we see simply such members of an infinite collection as happen to be the nearest our system? A few such questions as these we are perhaps beginning to answer; but hundreds, thousands, perhaps even millions, of years may elapse without our reaching a complete solution. Yet the astronomer does not view them as Kantian antinomies, in the nature of things insoluble, but as questions to which he may hopefully look for at least a partial answer.

The problem of the distances of the stars is of peculiar interest in connection with the Copernican system. The greatest objection to this system, which must have been more clearly seen by astronomers themselves than by any others, was found in the absence of any apparent parallax of the stars. If the earth performed such an immeasurable circle around the sun as Copernicus maintained, then, as it passed from side to side of its orbit, the stars outside the solar system must appear to have a corresponding motion in the other direction, and thus to swing back and forth as the earth moved in one and the other direction. The fact that not the slightest swing of that sort could be seen was, from the time of Ptolemy, the basis on which the doctrine of the earth's immobility rested. The difficulty was not grappled with by Copernicus or his immediate successors. The idea that Nature would not squander space by allowing immeasurable stretches of it to go unused seems to have been one from which medieval thinkers could not entirely break away. The consideration that there could be no need of any such economy, because the supply was infinite, might have been theoretically acknowledged, but was not practically felt. The fact is that magnificent as was the conception of Copernicus, it was dwarfed by the conception of stretches from star to star so vast that the whole orbit of the earth was only a point in comparison.

An indication of the extent to which the difficulty thus arising was felt is seen in the title of a book published by Horrebow, the Danish astronomer, some two centuries ago. This industrious observer, one of the first who used an instrument resembling our meridian transit of the present day, determined to see if he could find the parallax of the stars by observing the intervals at which a pair of stars in opposite quarters of the heavens crossed his meridian at opposite seasons of the year. When, as he thought, he had won success, he published his observations and conclusions under the title of Copernicus Triumphans. But alas! the keen criticism of his successors showed that what he supposed to be a swing of the stars from season to season arose from a minute variation in the rate of his clock, due to the different temperatures to which it was exposed during the day and the night. The measurement of the distance even of the nearest stars evaded astronomical research until Bessel and Struve arose in the early part of the present century.

On some aspects of the problem of the extent of the universe light is being thrown even now. Evidence is gradually accumulating which points to the probability that the successive orders of smaller and smaller stars, which our continually increasing telescopic power brings into view, are not situated at greater and greater distances, but that we actually see the boundary of our universe. This indication lends a peculiar interest to various questions growing out of the motions of the stars. Quite possibly the problem of these motions will be the great one of the future astronomer. Even now it suggests thoughts and questions of the most far-reaching character.

I have seldom felt a more delicious sense of repose than when crossing the ocean during the summer months I sought a place where I could lie alone on the deck, look up at the constellations, with Lyra near the zenith, and, while listening to the clank of the engine, try to calculate the hundreds of millions of years which would be required by our ship to reach the star a Lyrae, if she could continue her course in that direction without ever stopping. It is a striking example of how easily we may fail to realize our knowledge when I say that I have thought many a time how deliciously one might pass those hundred millions of years in a journey to the star a Lyrae, without its occurring to me that we are actually making that very journey at a speed compared with which the motion of a steamship is slow indeed. Through every year, every hour, every minute, of human history from the first appearance of man on the earth, from the era of the builders of the Pyramids, through the times of Caesar and Hannibal, through the period of every event that history records, not merely our earth, but the sun and the whole solar system with it, have been speeding their way towards the star of which I speak on a journey of which we know neither the beginning nor the end. We are at this moment thousands of miles nearer to a Lyrae than we were a few minutes ago when I began this discourse, and through every future moment, for untold thousands of years to come, the earth and all there is on it will be nearer to a Lyrae, or nearer to the place where that star now is, by hundreds of miles for every minute of time come and gone. When shall we get there? Probably in less than a million years, perhaps in half a million. We cannot tell exactly, but get there we must if the laws of nature and the laws of motion continue as they are. To attain to the stars was the seemingly vain wish of an ancient philosopher, but the whole human race is, in a certain sense, realizing this wish as rapidly as a speed of ten miles a second can bring it about.

I have called attention to this motion because it may, in the not distant future, afford the means of approximating to a solution of the problem already mentioned—that of the extent of the universe. Notwithstanding the success of astronomers during the present century in measuring the parallax of a number of stars, the most recent investigations show that there are very few, perhaps hardly more than a score, of stars of which the parallax, and therefore the distance, has been determined with any approach to certainty. Many parallaxes determined about the middle of the nineteenth century have had to disappear before the powerful tests applied by measures with the heliometer; others have been greatly reduced and the distances of the stars increased in proportion. So far as measurement goes, we can only say of the distances of all the stars, except the few whose parallaxes have been determined, that they are immeasurable. The radius of the earth's orbit, a line more than ninety millions of miles in length, not only vanishes from sight before we reach the distance of the great mass of stars, but becomes such a mere point that when magnified by the powerful instruments of modern times the most delicate appliances fail to make it measurable. Here the solar motion comes to our help. This motion, by which, as I have said, we are carried unceasingly through space, is made evident by a motion of most of the stars in the opposite direction, just as passing through a country on a railway we see the houses on the right and on the left being left behind us. It is clear enough that the apparent motion will be more rapid the nearer the object. We may therefore form some idea of the distance of the stars when we know the amount of the motion. It is found that in the great mass of stars of the sixth magnitude, the smallest visible to the naked eye, the motion is about three seconds per century. As a measure thus stated does not convey an accurate conception of magnitude to one not practised in the subject, I would say that in the heavens, to the ordinary eye, a pair of stars will appear single unless they are separated by a distance of 150 or 200 seconds. Let us, then, imagine ourselves looking at a star of the sixth magnitude, which is at rest while we are carried past it with the motion of six to eight miles per second which I have described. Mark its position in the heavens as we see it to-day; then let its position again be marked five thousand years hence. A good eye will just be able to perceive that there are two stars marked instead of one. The two would be so close together that no distinct space between them could be perceived by unaided vision. It is due to the magnifying power of the telescope, enlarging such small apparent distances, that the motion has been determined in so small a period as the one hundred and fifty years during which accurate observations of the stars have been made.

The motion just described has been fairly well determined for what, astronomically speaking, are the brighter stars; that is to say, those visible to the naked eye. But how is it with the millions of faint telescopic stars, especially those which form the cloud masses of the Milky Way? The distance of these stars is undoubtedly greater, and the apparent motion is therefore smaller. Accurate observations upon such stars have been commenced only recently, so that we have not yet had time to determine the amount of the motion. But the indication seems to be that it will prove quite a measurable quantity and that before the twentieth century has elapsed, it will be determined for very much smaller stars than those which have heretofore been studied. A photographic chart of the whole heavens is now being constructed by an association of observatories in some of the leading countries of the world. I cannot say all the leading countries, because then we should have to exclude our own, which, unhappily, has taken no part in this work. At the end of the twentieth century we may expect that the work will be repeated. Then, by comparing the charts, we shall see the effect of the solar motion and perhaps get new light upon the problem in question.

Closely connected with the problem of the extent of the universe is another which appears, for us, to be insoluble because it brings us face to face with infinity itself. We are familiar enough with eternity, or, let us say, the millions or hundreds of millions of years which geologists tell us must have passed while the crust of the earth was assuming its present form, our mountains being built, our rocks consolidated, and successive orders of animals coming and going. Hundreds of millions of years is indeed a long time, and yet, when we contemplate the changes supposed to have taken place during that time, we do not look out on eternity itself, which is veiled from our sight, as it were, by the unending succession of changes that mark the progress of time. But in the motions of the stars we are brought face to face with eternity and infinity, covered by no veil whatever. It would be bold to speak dogmatically on a subject where the springs of being are so far hidden from mortal eyes as in the depths of the universe. But, without declaring its positive certainty, it must be said that the conclusion seems unavoidable that a number of stars are moving with a speed such that the attraction of all the bodies of the universe could never stop them. One such case is that of Arcturus, the bright reddish star familiar to mankind since the days of Job, and visible near the zenith on the clear evenings of May and June. Yet another case is that of a star known in astronomical nomenclature as 1830 Groombridge, which exceeds all others in its angular proper motion as seen from the earth. We should naturally suppose that it seems to move so fast because it is near us. But the best measurements of its parallax seem to show that it can scarcely be less than two million times the distance of the earth from the sun, while it may be much greater. Accepting this result, its velocity cannot be much less than two hundred miles per second, and may be much more. With this speed it would make the circuit of our globe in two minutes, and had it gone round and round in our latitudes we should have seen it fly past us a number of times since I commenced this discourse. It would make the journey from the earth to the sun in five days. If it is now near the centre of our universe it would probably reach its confines in a million of years. So far as our knowledge goes, there is no force in nature which would ever have set it in motion and no force which can ever stop it. What, then, was the history of this star, and, if there are planets circulating around, what the experience of beings who may have lived on those planets during the ages which geologists and naturalists assure us our earth has existed? Was there a period when they saw at night only a black and starless heaven? Was there a time when in that heaven a small faint patch of light began gradually to appear? Did that patch of light grow larger and larger as million after million of years elapsed? Did it at last fill the heavens and break up into constellations as we now see them? As millions more of years elapse will the constellations gather together in the opposite quarter and gradually diminish to a patch of light as the star pursues its irresistible course of two hundred miles per second through the wilderness of space, leaving our universe farther and farther behind it, until it is lost in the distance? If the conceptions of modern science are to be considered as good for all time—a point on which I confess to a large measure of scepticism—then these questions must be answered in the affirmative.

The problems of which I have so far spoken are those of what may be called the older astronomy. If I apply this title it is because that branch of the science to which the spectroscope has given birth is often called the new astronomy. It is commonly to be expected that a new and vigorous form of scientific research will supersede that which is hoary with antiquity. But I am not willing to admit that such is the case with the old astronomy, if old we may call it. It is more pregnant with future discoveries today than it ever has been, and it is more disposed to welcome the spectroscope as a useful handmaid, which may help it on to new fields, than it is to give way to it. How useful it may thus become has been recently shown by a Dutch astronomer, who finds that the stars having one type of spectrum belong mostly to the Milky Way, and are farther from us than the others.

In the field of the newer astronomy perhaps the most interesting work is that associated with comets. It must be confessed, however, that the spectroscope has rather increased than diminished the mystery which, in some respects, surrounds the constitution of these bodies. The older astronomy has satisfactorily accounted for their appearance, and we might also say for their origin and their end, so far as questions of origin can come into the domain of science. It is now known that comets are not wanderers through the celestial spaces from star to star, but must always have belonged to our system. But their orbits are so very elongated that thousands, or even hundreds of thousands, of years are required for a revolution. Sometimes, however, a comet passing near to Jupiter is so fascinated by that planet that, in its vain attempts to follow it, it loses so much of its primitive velocity as to circulate around the sun in a period of a few years, and thus to become, apparently, a new member of our system. If the orbit of such a comet, or in fact of any comet, chances to intersect that of the earth, the latter in passing the point of intersection encounters minute particles which cause a meteoric shower.

But all this does not tell us much about the nature and make-up of a comet. Does it consist of nothing but isolated particles, or is there a solid nucleus, the attraction of which tends to keep the mass together? No one yet knows. The spectroscope, if we interpret its indications in the usual way, tells us that a comet is simply a mass of hydrocarbon vapor, shining by its own light. But there must be something wrong in this interpretation. That the light is reflected sunlight seems to follow necessarily from the increased brilliancy of the comet as it approaches the sun and its disappearance as it passes away.

Great attention has recently been bestowed upon the physical constitution of the planets and the changes which the surfaces of those bodies may undergo. In this department of research we must feel gratified by the energy of our countrymen who have entered upon it. Should I seek to even mention all the results thus made known I might be stepping on dangerous ground, as many questions are still unsettled. While every astronomer has entertained the highest admiration for the energy and enthusiasm shown by Mr. Percival Lowell in founding an observatory in regions where the planets can be studied under the most favorable conditions, they cannot lose sight of the fact that the ablest and most experienced observers are liable to error when they attempt to delineate the features of a body 50,000,000 or 100,000,000 miles away through such a disturbing medium as our atmosphere. Even on such a subject as the canals of Mars doubts may still be felt. That certain markings to which Schiaparelli gave the name of canals exist, few will question. But it may be questioned whether these markings are the fine, sharp, uniform lines found on Schiaparelli's map and delineated in Lowell's beautiful book. It is certainly curious that Barnard at Mount Hamilton, with the most powerful instrument and under the most favorable circumstances, does not see these markings as canals.

I can only mention among the problems of the spectroscope the elegant and remarkable solution of the mystery surrounding the rings of Saturn, which has been effected by Keeler at Allegheny. That these rings could not be solid has long been a conclusion of the laws of mechanics, but Keeler was the first to show that they really consist of separate particles, because the inner portions revolve more rapidly than the outer.

The question of the atmosphere of Mars has also received an important advance by the work of Campbell at Mount Hamilton. Although it is not proved that Mars has no atmosphere, for the existence of some atmosphere can scarcely be doubted, yet the Mount Hamilton astronomer seems to have shown, with great conclusiveness, that it is so rare as not to produce any sensible absorption of the solar rays.

I have left an important subject for the close. It belongs entirely to the older astronomy, and it is one with which I am glad to say this observatory is expected to especially concern itself. I refer to the question of the variation of latitudes, that singular phenomenon scarcely suspected ten years ago, but brought out by observations in Germany during the past eight years, and reduced to law with such brilliant success by our own Chandler. The north pole is not a fixed point on the earth's surface, but moves around in rather an irregular way. True, the motion is small; a circle of sixty feet in diameter will include the pole in its widest range. This is a very small matter so far as the interests of daily life are concerned; but it is very important to the astronomer. It is not simply a motion of the pole of the earth, but a wobbling of the solid earth itself. No one knows what conclusions of importance to our race may yet follow from a study of the stupendous forces necessary to produce even this slight motion.

The director of this new observatory has already distinguished himself in the delicate and difficult work of investigating this motion, and I am glad to know that he is continuing the work here with one of the finest instruments ever used for the purpose, a splendid product of American mechanical genius. I can assure you that astronomers the world over will look with the greatest interest for Professor Doolittle's success in the arduous task he has undertaken.

There is one question connected with these studies of the universe on which I have not touched, and which is, nevertheless, of transcendent interest. What sort of life, spiritual and intellectual, exists in distant worlds? We cannot for a moment suppose that our little planet is the only one throughout the whole universe on which may be found the fruits of civilization, family affection, friendship, the desire to penetrate the mysteries of creation. And yet this question is not to-day a problem of astronomy, nor can we see any prospect that it ever will be, for the simple reason that science affords us no hope of an answer to any question that we may send through the fathomless abyss. When the spectroscope was in its infancy it was suggested that possibly some difference might be found in the rays reflected from living matter, especially from vegetation, that might enable us to distinguish them from rays reflected by matter not endowed with life. But this hope has not been realized, nor does it seem possible to realize it. The astronomer cannot afford to waste his energies on hopeless speculation about matters of which he cannot learn anything, and he therefore leaves this question of the plurality of worlds to others who are as competent to discuss it as he is. All he can tell the world is:

He who through vast immensity can pierce,
See worlds on worlds compose one universe;
Observe how system into system runs,
What other planets circle other suns,
What varied being peoples every star,
May tell why Heaven has made us as we are.

ASPECTS OF AMERICAN ASTRONOMY

[Footnote: Address delivered at the University of Chicago, October 22, 1897, in connection with the dedication of the Yerkes Observatory. Printed in the Astro physical Journal. November, 1897.]

The University of Chicago yesterday accepted one of the most munificent gifts ever made for the promotion of any single science, and with appropriate ceremonies dedicated it to the increase of our knowledge of the heavenly bodies.

The president of your university has done me the honor of inviting me to supplement what was said on that occasion by some remarks of a more general nature suggested by the celebration. One is naturally disposed to say first what is uppermost in his mind. At the present moment this will naturally be the general impression made by what has been seen and heard. The ceremonies were attended, not only by a remarkable delegation of citizens, but by a number of visiting astronomers which seems large when we consider that the profession itself is not at all numerous in any country. As one of these, your guests, I am sure that I give expression only to their unanimous sentiment in saying that we have been extremely gratified in many ways by all that we have seen and heard. The mere fact of so munificent a gift to science cannot but excite universal admiration. We knew well enough that it was nothing more than might have been expected from the public spirit of this great West; but the first view of a towering snowpeak is none the less impressive because you have learned in your geography how many feet high it is, and great acts are none the less admirable because they correspond to what you have heard and read, and might therefore be led to expect.

The next gratifying feature is the great public interest excited by the occasion. That the opening of a purely scientific institution should have led so large an assemblage of citizens to devote an entire day, including a long journey by rail, to the celebration of yesterday is something most suggestive from its unfamiliarity. A great many scientific establishments have been inaugurated during the last half-century, but if on any such occasion so large a body of citizens has gone so great a distance to take part in the inauguration, the fact has at the moment escaped my mind.

That the interest thus shown is not confined to the hundreds of attendants, but must be shared by your great public, is shown by the unfailing barometer of journalism. Here we have a field in which the non-survival of the unfit is the rule in its most ruthless form. The journals that we see and read are merely the fortunate few of a countless number, dead and forgotten, that did not know what the public wanted to read about. The eagerness shown by the representatives of your press in recording everything your guests would say was accomplished by an enterprise in making known everything that occurred, and, in case of an emergency requiring a heroic measure, what did NOT occur, showing that smart journalists of the East must have learned their trade, or at least breathed their inspiration, in these regions. I think it was some twenty years since I told a European friend that the eighth wonder of the world was a Chicago daily newspaper. Since that time the course of journalistic enterprise has been in the reverse direction to that of the course of empire, eastward instead of westward.

It has been sometimes said—wrongfully, I think—that scientific men form a mutual admiration society. One feature of the occasion made me feel that we, your guests, ought then and there to have organized such a society and forthwith proceeded to business. This feature consisted in the conferences on almost every branch of astronomy by which the celebration of yesterday was preceded. The fact that beyond the acceptance of a graceful compliment I contributed nothing to these conferences relieves me from the charge of bias or self-assertion in saying that they gave me a new and most inspiring view of the energy now being expended in research by the younger generation of astronomers. All the experience of the past leads us to believe that this energy will reap the reward which nature always bestows upon those who seek her acquaintance from unselfish motives. In one way it might appear that little was to be learned from a meeting like that of the present week. Each astronomer may know by publications pertaining to the science what all the others are doing. But knowledge obtained in this way has a sort of abstractness about it a little like our knowledge of the progress of civilization in Japan, or of the great extent of the Australian continent. It was, therefore, a most happy thought on the part of your authorities to bring together the largest possible number of visiting astronomers from Europe, as well as America, in order that each might see, through the attrition of personal contact, what progress the others were making in their researches. To the visitors at least I am sure that the result of this meeting has been extremely gratifying. They earnestly hope, one and all, that the callers of the conference will not themselves be more disappointed in its results; that, however little they may have actually to learn of methods and results, they will feel stimulated to well-directed efforts and find themselves inspired by thoughts which, however familiar, will now be more easily worked out.

We may pass from the aspects of the case as seen by the strictly professional class to those general aspects fitted to excite the attention of the great public. From the point of view of the latter it may well appear that the most striking feature of the celebration is the great amount of effort which is shown to be devoted to the cultivation of a field quite outside the ordinary range of human interests. The workers whom we see around us are only a detachment from an army of investigators who, in many parts of the world, are seeking to explore the mysteries of creation. Why so great an expenditure of energy? Certainly not to gain wealth, for astronomy is perhaps the one field of scientific work which, in our expressive modern phrase, "has no money in it." It is true that the great practical use of astronomical science to the country and the world in affording us the means of determining positions on land and at sea is frequently pointed out. It is said that an Astronomer Royal of England once calculated that every meridian observation of the moon made at Greenwich was worth a pound sterling, on account of the help it would afford to the navigation of the ocean. An accurate map of the United States cannot be constructed without astronomical observations at numerous points scattered over the whole country, aided by data which great observatories have been accumulating for more than a century, and must continue to accumulate in the future.

But neither the measurement of the earth, the making of maps, nor the aid of the navigator is the main object which the astronomers of to-day have in view. If they do not quite share the sentiment of that eminent mathematician, who is said to have thanked God that his science was one which could not be prostituted to any useful purpose, they still know well that to keep utilitarian objects in view would only prove and handicap on their efforts. Consequently they never ask in what way their science is going to benefit mankind. As the great captain of industry is moved by the love of wealth, and the political leader by the love of power over men, so the astronomer is moved by the love of knowledge for its own sake, and not for the sake of its useful applications. Yet he is proud to know that his science has been worth more to mankind than it has cost. He does not value its results merely as a means of crossing the ocean or mapping the country, for he feels that man does not live by bread alone. If it is not more than bread to know the place we occupy in the universe, it is certainly something which we should place not far behind the means of subsistence.

That we now look upon a comet as something very interesting, of which the sight affords us a pleasure unmixed with fear of war, pestilence, or other calamity, and of which we therefore wish the return, is a gain we cannot measure by money. In all ages astronomy has been an index to the civilization of the people who cultivated it. It has been crude or exact, enlightened or mingled with superstition, according to the current mode of thought. When once men understand the relation of the planet on which they dwell to the universe at large, superstition is doomed to speedy extinction. This alone is an object worth more than money.

Astronomy may fairly claim to be that science which transcends all others in its demands upon the practical application of our reasoning powers. Look at the stars that stud the heavens on a clear evening. What more hopeless problem to one confined to earth than that of determining their varying distances, their motions, and their physical constitution? Everything on earth we can handle and investigate. But how investigate that which is ever beyond our reach, on which we can never make an experiment? On certain occasions we see the moon pass in front of the sun and hide it from our eyes. To an observer a few miles away the sun was not entirely hidden, for the shadow of the moon in a total eclipse is rarely one hundred miles wide. On another continent no eclipse at all may have been visible. Who shall take a map of the world and mark upon it the line on which the moon's shadow will travel during some eclipse a hundred years hence? Who shall map out the orbits of the heavenly bodies as they are going to appear in a hundred thousand years? How shall we ever know of what chemical elements the sun and the stars are made? All this has been done, but not by the intellect of any one man. The road to the stars has been opened only by the efforts of many generations of mathematicians and observers, each of whom began where his predecessor had left off.

We have reached a stage where we know much of the heavenly bodies. We have mapped out our solar system with great precision. But how with that great universe of millions of stars in which our solar system is only a speck of star-dust, a speck which a traveller through the wilds of space might pass a hundred times without notice? We have learned much about this universe, though our knowledge of it is still dim. We see it as a traveller on a mountain-top sees a distant city in a cloud of mist, by a few specks of glimmering light from steeples or roofs. We want to know more about it, its origin and its destiny; its limits in time and space, if it has any; what function it serves in the universal economy. The journey is long, yet we want, in knowledge at least, to make it. Hence we build observatories and train observers and investigators. Slow, indeed, is progress in the solution of the greatest of problems, when measured by what we want to know. Some questions may require centuries, others thousands of years for their answer. And yet never was progress more rapid than during our time. In some directions our astronomers of to-day are out of sight of those of fifty years ago; we are even gaining heights which twenty years ago looked hopeless. Never before had the astronomer so much work—good, hard, yet hopeful work—before him as to-day. He who is leaving the stage feels that he has only begun and must leave his successors with more to do than his predecessors left him.

To us an interesting feature of this progress is the part taken in it by our own country. The science of our day, it is true, is of no country. Yet we very appropriately speak of American science from the fact that our traditional reputation has not been that of a people deeply interested in the higher branches of intellectual work. Men yet living can remember when in the eyes of the universal church of learning, all cisatlantic countries, our own included, were partes infidelium.

Yet American astronomy is not entirely of our generation. In the middle of the last century Professor Winthrop, of Harvard, was an industrious observer of eclipses and kindred phenomena, whose work was recorded in the transactions of learned societies. But the greatest astronomical activity during our colonial period was that called out by the transit of Venus in 1769, which was visible in this country. A committee of the American Philosophical Society, at Philadelphia, organized an excellent system of observations, which we now know to have been fully as successful, perhaps more so, than the majority of those made on other continents, owing mainly to the advantages of air and climate. Among the observers was the celebrated Rittenhouse, to whom is due the distinction of having been the first American astronomer whose work has an important place in the history of the science. In addition to the observations which he has left us, he was the first inventor or proposer of the collimating telescope, an instrument which has become almost a necessity wherever accurate observations are made. The fact that the subsequent invention by Bessel may have been independent does not detract from the merits of either.

Shortly after the transit of Venus, which I have mentioned, the war of the Revolution commenced. The generation which carried on that war and the following one, which framed our Constitution and laid the bases of our political institutions, were naturally too much occupied with these great problems to pay much attention to pure science. While the great mathematical astronomers of Europe were laying the foundation of celestial mechanics their writings were a sealed book to every one on this side of the Atlantic, and so remained until Bowditch appeared, early in the present century. His translation of the Mecanique Celeste made an epoch in American science by bringing the great work of Laplace down to the reach of the best American students of his time.

American astronomers must always honor the names of Rittenhouse and Bowditch. And yet in one respect their work was disappointing of results.

Neither of them was the founder of a school. Rittenhouse left no successor to carry on his work. The help which Bowditch afforded his generation was invaluable to isolated students who, here and there, dived alone and unaided into the mysteries of the celestial motions. His work was not mainly in the field of observational astronomy, and therefore did not materially influence that branch of science. In 1832 Professor Airy, afterwards Astronomer Royal of England, made a report to the British Association on the condition of practical astronomy in various countries. In this report he remarked that he was unable to say anything about American astronomy because, so far as he knew, no public observatory existed in the United States.

William C. Bond, afterwards famous as the first director of the Harvard Observatory, was at that time making observations with a small telescope, first near Boston and afterwards at Cambridge. But with so meagre an outfit his establishment could scarcely lay claim to being an astronomical observatory, and it was not surprising if Airy did not know anything of his modest efforts.

If at this time Professor Airy had extended his investigations into yet another field, with a view of determining the prospects for a great city at the site of Fort Dearborn, on the southern shore of Lake Michigan, he would have seen as little prospect of civic growth in that region as of a great development of astronomy in the United States at large. A plat of the proposed town of Chicago had been prepared two years before, when the place contained perhaps half a dozen families. In the same month in which Professor Airy made his report, August, 1832, the people of the place, then numbering twenty-eight voters, decided to become incorporated, and selected five trustees to carry on their government.

In 1837 a city charter was obtained from the legislature of Illinois. The growth of this infant city, then small even for an infant, into the great commercial metropolis of the West has been the just pride of its people and the wonder of the world. I mention it now because of a remarkable coincidence. With this civic growth has quietly gone on another, little noted by the great world, and yet in its way equally wonderful and equally gratifying to the pride of those who measure greatness by intellectual progress. Taking knowledge of the universe as a measure of progress, I wish to invite attention to the fact that American astronomy began with your city, and has slowly but surely kept pace with it, until to-day our country stands second only to Germany in the number of researches being prosecuted, and second to none in the number of men who have gained the highest recognition by their labors.

In 1836 Professor Albert Hopkins, of Williams College, and Professor Elias Loomis, of Western Reserve College, Ohio, both commenced little observatories. Professor Loomis went to Europe for all his instruments, but

Hopkins was able even then to get some of his in this country. Shortly afterwards a little wooden structure was erected by Captain Gilliss on Capitol Hill, at Washington, and supplied with a transit instrument for observing moon culminations, in conjunction with Captain Wilkes, who was then setting out on his exploring expedition to the southern hemisphere. The date of these observatories was practically the same as that on which a charter for the city of Chicago was obtained from the legislature. With their establishment the population of your city had increased to 703.

The next decade, 1840 to 1850, was that in which our practical astronomy seriously commenced. The little observatory of Captain Gilliss was replaced by the Naval, then called the National Observatory, erected at Washington during the years 1843-44, and fitted out with what were then the most approved instruments. About the same time the appearance of the great comet of 1843 led the citizens of Boston to erect the observatory of Harvard College. Thus it is little more than a half-century since the two principal observatories in the United States were established. But we must not for a moment suppose that the mere erection of an observatory can mark an epoch in scientific history. What must make the decade of which I speak ever memorable in American astronomy was not merely the erection of buildings, but the character of the work done by astronomers away from them as well as in them.

The National Observatory soon became famous by two remarkable steps which raised our country to an important position among those applying modern science to practical uses. One of these consisted of the researches of Sears Cook Walker on the motion of the newly discovered planet Neptune. He was the first astronomer to determine fairly good elements of the orbit of that planet, and, what is yet more remarkable, he was able to trace back the movement of the planet in the heavens for half a century and to show that it had been observed as a fixed star by Lalande in 1795, without the observer having any suspicion of the true character of the object.

The other work to which I refer was the application to astronomy and to the determination of longitudes of the chronographic method of registering transits of stars or other phenomena requiring an exact record of the instant of their occurrence. It is to be regretted that the history of this application has not been fully written. In some points there seems to be as much obscurity as with the discovery of ether as an anaesthetic, which took place about the same time. Happily, no such contest has been fought over the astronomical as over the surgical discovery, the fact being that all who were engaged in the application of the new method were more anxious to perfect it than they were to get credit for themselves. We know that Saxton, of the Coast Survey; Mitchell and Locke, of Cincinnati; Bond, at Cambridge, as well as Walker, and other astronomers at the Naval Observatory, all worked at the apparatus; that Maury seconded their efforts with untiring zeal; that it was used to determine the longitude of Baltimore as early as 1844 by Captain Wilkes, and that it was put into practical use in recording observations at the Naval Observatory as early as 1846.

At the Cambridge Observatory the two Bonds, father and son, speedily began to show the stuff of which the astronomer is made. A well-devised system of observations was put in operation. The discovery of the dark ring of Saturn and of a new satellite to that planet gave additional fame to the establishment.

Nor was activity confined to the observational side of the science. The same decade of which I speak was marked by the beginning of Professor Pierce's mathematical work, especially his determination of the perturbations of Uranus and Neptune. At this time commenced the work of Dr. B. A. Gould, who soon became the leading figure in American astronomy. Immediately on graduating at Harvard in 1845, he determined to devote all the energies of his life to the prosecution of his favorite science. He studied in Europe for three years, took the doctor's degree at Gottingen, came home, founded the Astronomical Journal, and took an active part in that branch of the work of the Coast Survey which included the determination of longitudes by astronomical methods.

An episode which may not belong to the history of astronomy must be acknowledged to have had a powerful influence in exciting public interest in that science. Professor O. M. Mitchell, the founder and first director of the Cincinnati Observatory, made the masses of our intelligent people acquainted with the leading facts of astronomy by courses of lectures which, in lucidity and eloquence, have never been excelled. The immediate object of the lectures was to raise funds for establishing his observatory and fitting it out with a fine telescope. The popular interest thus excited in the science had an important effect in leading the public to support astronomical research. If public support, based on public interest, is what has made the present fabric of American astronomy possible, then should we honor the name of a man whose enthusiasm leavened the masses of his countrymen with interest in our science.

The Civil War naturally exerted a depressing influence upon our scientific activity. The cultivator of knowledge is no less patriotic than his fellow-citizens, and vies with them in devotion to the public welfare. The active interest which such cultivators took, first in the prosecution of the war and then in the restoration of the Union, naturally distracted their attention from their favorite pursuits. But no sooner was political stability reached than a wave of intellectual activity set in, which has gone on increasing up to the present time. If it be true that never before in our history has so much attention been given to education as now; that never before did so many men devote themselves to the diffusion of knowledge, it is no less true that never was astronomical work so energetically pursued among us as at the present time.

One deplorable result of the Civil War was that Gould's Astronomical Journal had to be suspended. Shortly after the restoration of peace, instead of re-establishing the journal, its founder conceived the project of exploring the southern heavens. The northern hemisphere being the seat of civilization, that portion of the sky which could not be seen from our latitudes was comparatively neglected. What had been done in the southern hemisphere was mostly the occasional work of individuals and of one or two permanent observatories. The latter were so few in number and so meagre in their outfit that a splendid field was open to the inquirer. Gould found the patron which he desired in the government of the Argentine Republic, on whose territory he erected what must rank in the future as one of the memorable astronomical establishments of the world. His work affords a most striking example of the principle that the astronomer is more important than his instruments. Not only were the means at the command of the Argentine Observatory slender in the extreme when compared with those of the favored institutions of the North, but, from the very nature of the case, the Argentine Republic could not supply trained astronomers. The difficulties thus growing out of the administration cannot be overestimated. And yet the sixteen great volumes in which the work of the institution has been published will rank in the future among the classics of astronomy.

Another wonderful focus of activity, in which one hardly knows whether he ought most to admire the exhaustless energy or the admirable ingenuity which he finds displayed, is the Harvard Observatory. Its work has been aided by gifts which have no parallel in the liberality that prompted them. Yet without energy and skill such gifts would have been useless. The activity of the establishment includes both hemispheres. Time would fail to tell how it has not only mapped out important regions of the heavens from the north to the south pole, but analyzed the rays of light which come from hundreds of thousands of stars by recording their spectra in permanence on photographic plates.

The work of the establishment is so organized that a new star cannot appear in any part of the heavens nor a known star undergo any noteworthy change without immediate detection by the photographic eye of one or more little telescopes, all-seeing and never-sleeping policemen that scan the heavens unceasingly while the astronomer may sleep, and report in the morning every case of irregularity in the proceedings of the heavenly bodies.

Yet another example, showing what great results may be obtained with limited means, is afforded by the Lick Observatory, on Mount Hamilton, California. During the ten years of its activity its astronomers have made it known the world over by works and discoveries too varied and numerous to be even mentioned at the present time.

The astronomical work of which I have thus far spoken has been almost entirely that done at observatories. I fear that I may in this way have strengthened an erroneous impression that the seat of important astronomical work is necessarily connected with an observatory. It must be admitted that an institution which has a local habitation and a magnificent building commands public attention so strongly that valuable work done elsewhere may be overlooked. A very important part of astronomical work is done away from telescopes and meridian circles and requires nothing but a good library for its prosecution. One who is devoted to this side of the subject may often feel that the public does not appreciate his work at its true relative value from the very fact that he has no great buildings or fine instruments to show. I may therefore be allowed to claim as an important factor in the American astronomy of the last half-century an institution of which few have heard and which has been overlooked because there was nothing about it to excite attention.

In 1849 the American Nautical Almanac office was established by a Congressional appropriation. The title of this publication is somewhat misleading in suggesting a simple enlargement of the family almanac which the sailor is to hang up in his cabin for daily use. The fact is that what started more than a century ago as a nautical almanac has since grown into an astronomical ephemeris for the publication of everything pertaining to times, seasons, eclipses, and the motions of the heavenly bodies. It is the work in which astronomical observations made in all the great observatories of the world are ultimately utilized for scientific and public purposes. Each of the leading nations of western Europe issues such a publication. When the preparation and publication of the American ephemeris was decided upon the office was first established in Cambridge, the seat of Harvard University, because there could most readily be secured the technical knowledge of mathematics and theoretical astronomy necessary for the work.

A field of activity was thus opened, of which a number of able young men who have since earned distinction in various walks of life availed themselves. The head of the office, Commander Davis, adopted a policy well fitted to promote their development. He translated the classic work of Gauss, Theoria Motus Corporum Celestium, and made the office a sort of informal school, not, indeed, of the modern type, but rather more like the classic grove of Hellas, where philosophers conducted their discussions and profited by mutual attrition. When, after a few years of experience, methods were well established and a routine adopted, the office was removed to Washington, where it has since remained. The work of preparing the ephemeris has, with experience, been reduced to a matter of routine which may be continued indefinitely, with occasional changes in methods and data, and improvements to meet the increasing wants of investigators.

The mere preparation of the ephemeris includes but a small part of the work of mathematical calculation and investigation required in astronomy. One of the great wants of the science to-day is the reduction of the observations made during the first half of the present century, and even during the last half of the preceding one. The labor which could profitably be devoted to this work would be more than that required in any one astronomical observatory. It is unfortunate for this work that a great building is not required for its prosecution because its needfulness is thus very generally overlooked by that portion of the public interested in the progress of science. An organization especially devoted to it is one of the scientific needs of our time.

In such an epoch-making age as the present it is dangerous to cite any one step as making a new epoch. Yet it may be that when the historian of the future reviews the science of our day he will find the most remarkable feature of the astronomy of the last twenty years of our century to be the discovery that this steadfast earth of which the poets have told us is not, after all, quite steadfast; that the north and south poles move about a very little, describing curves so complicated that they have not yet been fully marked out. The periodic variations of latitude thus brought about were first suspected about 1880, and announced with some modest assurance by Kustner, of Berlin, a few years later. The progress of the views of astronomical opinion from incredulity to confidence was extremely slow until, about 1890, Chandler, of the United States, by an exhaustive discussion of innumerable results of observations, showed that the latitude of every point on the earth was subject to a double oscillation, one having a period of a year, the other of four hundred and twenty-seven days.

Notwithstanding the remarkable parallel between the growth of American astronomy and that of your city, one cannot but fear that if a foreign observer had been asked only half a dozen years ago at what point in the United States a great school of theoretical and practical astronomy, aided by an establishment for the exploration of the heavens, was likely to be established by the munificence of private citizens, he would have been wiser than most foreigners had he guessed Chicago. Had this place been suggested to him, I fear he would have replied that were it possible to utilize celestial knowledge in acquiring earthly wealth, here would be the most promising seat for such a school. But he would need to have been a little wiser than his generation to reflect that wealth is at the base of all progress in knowledge and the liberal arts; that it is only when men are relieved from the necessity of devoting all their energies to the immediate wants of life that they can lead the intellectual life, and that we should therefore look to the most enterprising commercial centre as the likeliest seat for a great scientific institution.

Now we have the school, and we have the observatory, which we hope will in the near future do work that will cast lustre on the name of its founder as well as on the astronomers who may be associated with it. You will, I am sure, pardon me if I make some suggestions on the subject of the future needs of the establishment. We want this newly founded institution to be a great success, to do work which shall show that the intellectual productiveness of your community will not be allowed to lag behind its material growth The public is very apt to feel that when some munificent patron of science has mounted a great telescope under a suitable dome, and supplied all the apparatus which the astronomer wants to use, success is assured. But such is not the case. The most important requisite, one more difficult to command than telescopes or observatories, may still be wanting. A great telescope is of no use without a man at the end of it, and what the telescope may do depends more upon this appendage than upon the instrument itself. The place which telescopes and observatories have taken in astronomical history are by no means proportional to their dimensions. Many a great instrument has been a mere toy in the hands of its owner. Many a small one has become famous.

Twenty years ago there was here in your own city a modest little instrument which, judged by its size, could not hold up its head with the great ones even of that day. It was the private property of a young man holding no scientific position and scarcely known to the public. And yet that little telescope is to-day among the famous ones of the world, having made memorable advances in the astronomy of double stars, and shown its owner to be a worthy successor of the Herschels and Struves in that line of work.

A hundred observers might have used the appliances of the Lick Observatory for a whole generation without finding the fifth satellite of Jupiter; without successfully photographing the cloud forms of the Milky Way; without discovering the extraordinary patches of nebulous light, nearly or quite invisible to the human eye, which fill some regions of the heavens.

When I was in Zurich last year I paid a visit to the little, but not unknown, observatory of its famous polytechnic school. The professor of astronomy was especially interested in the observations of the sun with the aid of the spectroscope, and among the ingenious devices which he described, not the least interesting was the method of photographing the sun by special rays of the spectrum, which had been worked out at the Kenwood Observatory in Chicago. The Kenwood Observatory is not, I believe, in the eye of the public, one of the noteworthy institutions of your city which every visitor is taken to see, and yet this invention has given it an important place in the science of our day.

Should you ask me what are the most hopeful features in the great establishment which you are now dedicating, I would say that they are not alone to be found in the size of your unequalled telescope, nor in the cost of the outfit, but in the fact that your authorities have shown their appreciation of the requirements of success by adding to the material outfit of the establishment the three men whose works I have described.

Gentlemen of the trustees, allow me to commend to your fostering care the men at the end of the telescope. The constitution of the astronomer shows curious and interesting features. If he is destined to advance the science by works of real genius, he must, like the poet, be born, not made. The born astronomer, when placed in command of a telescope, goes about using it as naturally and effectively as the babe avails itself of its mother's breast. He sees intuitively what less gifted men have to learn by long study and tedious experiment. He is moved to celestial knowledge by a passion which dominates his nature. He can no more avoid doing astronomical work, whether in the line of observations or research, than a poet can chain his Pegasus to earth. I do not mean by this that education and training will be of no use to him. They will certainly accelerate his early progress. If he is to become great on the mathematical side, not only must his genius have a bend in that direction, but he must have the means of pursuing his studies. And yet I have seen so many failures of men who had the best instruction, and so many successes of men who scarcely learned anything of their teachers, that I sometimes ask whether the great American celestial mechanician of the twentieth century will be a graduate of a university or of the backwoods.

Is the man thus moved to the exploration of nature by an unconquerable passion more to be envied or pitied? In no other pursuit does success come with such certainty to him who deserves it. No life is so enjoyable as that whose energies are devoted to following out the inborn impulses of one's nature. The investigator of truth is little subject to the disappointments which await the ambitious man in other fields of activity. It is pleasant to be one of a brotherhood extending over the world, in which no rivalry exists except that which comes out of trying to do better work than any one else, while mutual admiration stifles jealousy. And yet, with all these advantages, the experience of the astronomer may have its dark side. As he sees his field widening faster than he can advance he is impressed with the littleness of all that can be done in one short life. He feels the same want of successors to pursue his work that the founder of a dynasty may feel for heirs to occupy his throne. He has no desire to figure in history as a Napoleon of science whose conquests must terminate with his life. Even during his active career his work may be such a kind as to require the co-operation of others and the active support of the public. If he is disappointed in commanding these requirements, if he finds neither co-operation nor support, if some great scheme to which he may have devoted much of his life thus proves to be only a castle in the air, he may feel that nature has dealt hardly with him in not endowing him with passions like to those of other men.

In treating a theme of perennial interest one naturally tries to fancy what the future may have in store If the traveller, contemplating the ruins of some ancient city which in the long ago teemed with the life and activities of generations of men, sees every stone instinct with emotion and the dust alive with memories of the past, may he not be similarly impressed when he feels that he is looking around upon a seat of future empire—a region where generations yet unborn may take a leading part in moulding the history of the world? What may we not expect of that energy which in sixty years has transformed a straggling village into one of the world's great centres of commerce? May it not exercise a powerful influence on the destiny not only of the country but of the world? If so, shall the power thus to be exercised prove an agent of beneficence, diffusing light and life among nations, or shall it be the opposite?

The time must come ere long when wealth shall outgrow the field in which it can be profitably employed. In what direction shall its possessors then look? Shall they train a posterity which will so use its power as to make the world better that it has lived in it? Will the future heir to great wealth prefer the intellectual life to the life of pleasure?

We can have no more hopeful answer to these questions than the establishment of this great university in the very focus of the commercial activity of the West. Its connection with the institution we have been dedicating suggests some thoughts on science as a factor in that scheme of education best adapted to make the power of a wealthy community a benefit to the race at large. When we see what a factor science has been in our present civilization, how it has transformed the world and increased the means of human enjoyment by enabling men to apply the powers of nature to their own uses, it is not wonderful that it should claim the place in education hitherto held by classical studies. In the contest which has thus arisen I take no part but that of a peace-maker, holding that it is as important to us to keep in touch with the traditions of our race, and to cherish the thoughts which have come down to us through the centuries, as it is to enjoy and utilize what the present has to offer us. Speaking from this point of view, I would point out the error of making the utilitarian applications of knowledge the main object in its pursuit. It is an historic fact that abstract science—science pursued without any utilitarian end—has been at the base of our progress in the utilization of knowledge. If in the last century such men as Galvani and Volta had been moved by any other motive than love of penetrating the secrets of nature they would never have pursued the seemingly useless experiments they did, and the foundation of electrical science would not have been laid. Our present applications of electricity did not become possible until Ohm's mathematical laws of the electric current, which when first made known seemed little more than mathematical curiosities, had become the common property of inventors. Professional pride on the part of our own Henry led him, after making the discoveries which rendered the telegraph possible, to go no further in their application, and to live and die without receiving a dollar of the millions which the country has won through his agency.

In the spirit of scientific progress thus shown we have patriotism in its highest form—a sentiment which does not seek to benefit the country at the expense of the world, but to benefit the world by means of one's country. Science has its competition, as keen as that which is the life of commerce. But its rivalries are over the question who shall contribute the most and the best to the sum total of knowledge; who shall give the most, not who shall take the most. Its animating spirit is love of truth. Its pride is to do the greatest good to the greatest number. It embraces not only the whole human race but all nature in its scope. The public spirit of which this city is the focus has made the desert blossom as the rose, and benefited humanity by the diffusion of the material products of the earth. Should you ask me how it is in the future to use its influence for the benefit of humanity at large, I would say, look at the work now going on in these precincts, and study its spirit. Here are the agencies which will make "the voice of law the harmony of the world." Here is the love of country blended with love of the race. Here the love of knowledge is as unconfined as your commercial enterprise. Let not your youth come hither merely to learn the forms of vertebrates and the properties of oxides, but rather to imbibe that catholic spirit which, animating their growing energies, shall make the power they are to wield an agent of beneficence to all mankind.

THE UNIVERSE AS AN ORGANISM

[Footnote: Address before the Astronomical and Astrophysical Society of America, December 29, 1902]

If I were called upon to convey, within the compass of a single sentence, an idea of the trend of recent astronomical and physical science, I should say that it was in the direction of showing the universe to be a connected whole. The farther we advance in knowledge, the clearer it becomes that the bodies which are scattered through the celestial spaces are not completely independent existences, but have, with all their infinite diversity, many attributes in common.

In this we are going in the direction of certain ideas of the ancients which modern discovery long seemed to have contradicted. In the infancy of the race, the idea that the heavens were simply an enlarged and diversified earth, peopled by beings who could roam at pleasure from one extreme to the other, was a quite natural one. The crystalline sphere or spheres which contained all formed a combination of machinery revolving on a single plan. But all bonds of unity between the stars began to be weakened when Copernicus showed that there were no spheres, that the planets were isolated bodies, and that the stars were vastly more distant than the planets. As discovery went on and our conceptions of the universe were enlarged, it was found that the system of the fixed stars was made up of bodies so vastly distant and so completely isolated that it was difficult to conceive of them as standing in any definable relation to one another. It is true that they all emitted light, else we could not see them, and the theory of gravitation, if extended to such distances, a fact not then proved, showed that they acted on one another by their mutual gravitation. But this was all. Leaving out light and gravitation, the universe was still, in the time of Herschel, composed of bodies which, for the most part, could not stand in any

191

known relation one to the other.

When, forty years ago, the spectroscope was applied to analyze the light coming from the stars, a field was opened not less fruitful than that which the telescope made known to Galileo. The first conclusion reached was that the sun was composed almost entirely of the same elements that existed upon the earth. Yet, as the bodies of our solar system were evidently closely related, this was not remarkable. But very soon the same conclusion was, to a limited extent, extended to the fixed stars in general. Such elements as iron, hydrogen, and calcium were found not to belong merely to our earth, but to form important constituents of the whole universe. We can conceive of no reason why, out of the infinite number of combinations which might make up a spectrum, there should not be a separate kind of matter for each combination. So far as we know, the elements might merge into one another by insensible gradations. It is, therefore, a remarkable and suggestive fact when we find that the elements which make up bodies so widely separate that we can hardly imagine them having anything in common, should be so much the same.

In recent times what we may regard as a new branch of astronomical science is being developed, showing a tendency towards unity of structure throughout the whole domain of the stars. This is what we now call the science of stellar statistics. The very conception of such a science might almost appall us by its immensity. The widest statistical field in other branches of research is that occupied by sociology. Every country has its census, in which the individual inhabitants are classified on the largest scale and the combination of these statistics for different countries may be said to include all the interest of the human race within its scope. Yet this field is necessarily confined to the surface of our planet. In the field of stellar statistics millions of stars are classified as if each taken individually were of no more weight in the scale than a single inhabitant of China in the scale of the sociologist. And yet the most insignificant of these suns may, for aught we know, have planets revolving around it, the interests of whose inhabitants cover as wide a range as ours do upon our own globe.

The statistics of the stars may be said to have commenced with Herschel's gauges of the heavens, which were continued from time to time by various observers, never, however, on the largest scale. The subject was first opened out into an illimitable field of research through a paper presented by Kapteyn to the Amsterdam Academy of Sciences in 1893. The capital results of this paper were that different regions of space contain different kinds of stars and, more especially, that the stars of the Milky Way belong, in part at least, to a different class from those existing elsewhere. Stars not belonging to the Milky Way are, in large part, of a distinctly different class.

The outcome of Kapteyn's conclusions is that we are able to describe the universe as a single object, with some characters of an organized whole. A large part of the stars which compose it may be considered as divisible into two groups. One of these comprises the stars composing the great girdle of the Milky Way. These are distinguished from the others by being bluer in color, generally greater in absolute brilliancy, and affected, there is some reason to believe, with rather slower proper motions The other classes are stars with a greater or less shade of yellow in their color, scattered through a spherical space of unknown dimensions, but concentric with the Milky Way. Thus a sphere with a girdle passing around it forms the nearest approach to a conception of the universe which we can reach to-day. The number of stars in the girdle is much greater than that in the sphere.

The feature of the universe which should therefore command our attention is the arrangement of a large part of the stars which compose it in a ring, seemingly alike in all its parts, so far as general features are concerned. So far as research has yet gone, we are not able to say decisively that one region of this ring differs essentially from another. It may, therefore, be regarded as forming a structure built on a uniform plan throughout.

All scientific conclusions drawn from statistical data require a critical investigation of the basis on which they rest. If we are going, from merely counting the stars, observing their magnitudes and determining their proper motions, to draw conclusions as to the structure of the universe in space, the question may arise how we can form any estimate whatever of the possible distance of the stars, a conclusion as to which must be the very first step we take. We can hardly say that the parallaxes of more than one hundred stars have been measured with any approach to certainty. The individuals of this one hundred are situated at very different distances from us. We hope, by long and repeated observations, to make a fairly approximate determination of the parallaxes of all the stars whose distance is less than twenty times that of a Centauri. But how can we know anything about the distance of stars outside this sphere? What can we say against the view of Kepler that the space around our sun is very much thinner in stars than it is at a greater distance; in fact, that, the great mass of the stars may be situated between the surfaces of two concentrated spheres not very different in radius. May not this universe of stars be somewhat in the nature of a hollow sphere?

This objection requires very careful consideration on the part of all who draw conclusions as to the distribution of stars in space and as to the extent of the visible universe. The steps to a conclusion on the subject are briefly these: First, we have a general conclusion, the basis of which I have already set forth, that, to use a loose expression, there are likenesses throughout the whole diameter of the universe. There is, therefore, no reason to suppose that the region in which our system is situated differs in any essential degree from any other region near the central portion. Again, spectroscopic examinations seem to show that all the stars are in motion, and that we cannot say that those in one part of the universe move more rapidly than those in another. This result is of the greatest value for our purpose, because, when we consider only the apparent motions, as ordinarily observed, these are necessarily dependent upon the distance of the star. We cannot, therefore, infer the actual speed of a star from ordinary observations until we know its distance. But the results of spectroscopic measurements of radial velocity are independent of the distance of the star.

But let us not claim too much. We cannot yet say with certainty that the stars which form the agglomerations of the Milky Way have, beyond doubt, the same average motion as the stars in other regions of the universe. The difficulty is that these stars appear to us so faint individually, that the investigation of their spectra is still beyond the powers of our instruments. But the extraordinary feat performed at the Lick Observatory of measuring the radial motion of 1830 Groombridge, a star quite invisible to the naked eye, and showing that it is approaching our system with a speed of between fifty and sixty miles a second, may lead us to hope for a speedy solution of this question. But we need not await this result in order to reach very probable conclusions. The general outcome of researches on proper motions tends to strengthen the conclusions that the Keplerian sphere, if I may use this expression, has no very well marked existence. The laws of stellar velocity and the statistics of proper motions, while giving some color to the view that the space in which we are situated is thinner in stars than elsewhere, yet show that, as a general rule, there are no great agglomerations of stars elsewhere than in the region of the Milky Way.

With unity there is always diversity; in fact, the unity of the universe on which I have been insisting consists in part of diversity. It is very curious that, among the many thousands of stars which have been spectroscopically examined, no two are known to have absolutely the same physical constitution. It is true that there are a great many resemblances. Alpha Centauri, our nearest neighbor, if we can use such a word as "near" in speaking of its distance, has a spectrum very like that of our sun, and so has Capella. But even in these cases careful examination shows differences. These differences arise from variety in the combinations and temperature of the substances of which the star is made up. Quite likely also, elements not known on the earth may exist on the stars, but this is a point on which we cannot yet speak with certainty.

Perhaps the attribute in which the stars show the greatest variety is that of absolute luminosity. One hundred years ago it was naturally supposed that the brighter stars were the nearest to us, and this is doubtless true when we take the general average. But it was soon found that we cannot conclude that because a star is bright, therefore it is near. The most striking example of this is afforded by the absence of measurable parallaxes in the two bright stars, Canopus and Rigel, showing that these stars, though of the first magnitude, are immeasurably distant. A remarkable fact is that these conclusions coincide with that which we draw from the minuteness of the proper motions. Rigel has no motion that has certainly been shown by more than a century of observation, and it is not certain that Canopus has either. From this alone we may conclude, with a high degree of probability, that the distance of each is immeasurably great. We may say with certainty that the brightness of each is thousands of times that of the sun, and with a high degree of probability that it is hundreds of thousands of times. On the other hand, there are stars comparatively near us of which the light is not the hundredth part of the sun.

The universe may be a unit in two ways. One is that unity of structure to which our attention has just been directed. This might subsist forever without one body influencing another. The other form of unity leads us to view the universe as an organism. It is such by mutual action going on between its bodies. A few years ago we could hardly suppose or imagine that any other agents than gravitation and light could possibly pass through spaces so immense as those which separate the stars.

The most remarkable and hopeful characteristic of the unity of the universe is the evidence which is being gathered that there are other agencies whose exact nature is yet unknown to us, but which do pass from one heavenly body to another. The best established example of this yet obtained is afforded in the case of the sun and the earth.

The fact that the frequency of magnetic storms goes through a period of about eleven years, and is proportional to the frequency of sun-spots, has been well established. The recent work of Professor Bigelow shows the coincidence to be of remarkable exactness, the curves of the two phenomena being practically coincident so far as their general features are concerned. The conclusion is that spots on the sun and magnetic storms are due to the same cause. This cause cannot be any change in the ordinary radiation of the sun, because the best records of temperature show that, to whatever variations the sun's radiation may be subjected, they do not change in the period of the sun-spots. To appreciate the relation, we must recall that the researches of Hale with the spectro-heliograph show that spots are not the primary phenomenon of solar activity, but are simply the outcome of processes going on constantly in the sun which result in spots only in special regions and on special occasions. It does not, therefore, necessarily follow that a spot does cause a magnetic storm. What we should conclude is that the solar activity which produces a spot also produces the magnetic storm.

When we inquire into the possible nature of these relations between solar activity and terrestrial magnetism, we find ourselves so completely in the dark that the question of what is really proved by the coincidence may arise. Perhaps the most obvious explanation of fluctuations in the earth's magnetic field to be inquired into would be based on the hypothesis that the space through which the earth is moving is in itself a varying magnetic field of vast extent. This explanation is tested by inquiring whether the fluctuations in question can be explained by supposing a disturbing force which acts substantially in the same direction all over the globe. But a very obvious test shows that this explanation is untenable. Were it the correct one, the intensity of the force in some regions of the earth would be diminished and in regions where the needle pointed in the opposite direction would be increased in exactly the same degree. But there is no relation traceable either in any of the regular fluctuations of the magnetic force, or in those irregular ones which occur during a magnetic storm. If the horizontal force is increased in one part of the earth, it is very apt to show a simultaneous increase the world over, regardless of the direction in which the needle may point in various localities. It is hardly necessary to add that none of the fluctuations in terrestrial magnetism can be explained on the hypothesis that either the moon or the sun acts as a magnet. In such a case the action would be substantially in the same direction at the same moment the world over.

Such being the case, the question may arise whether the action producing a magnetic storm comes from the sun at all, and whether the fluctuations in the sun's activity, and in the earth's magnetic field may not be due to some cause external to both. All we can say in reply to this is that every effort to find such a cause has failed and that it is hardly possible to imagine any cause producing such an effect. It is true that the solar spots were, not many years ago, supposed to be due in some way to the action of the planets. But, for reasons which it would be tedious to go into at present, we may fairly regard this hypothesis as being completely disproved. There can, I conclude, be little doubt that the eleven-year cycle of change in the solar spots is due to a cycle going on in the sun itself. Such being the case, the corresponding change in the earth's magnetism must be due to the same cause.

We may, therefore, regard it as a fact sufficiently established to merit further investigation that there does emanate from the sun, in an irregular way, some agency adequate to produce a measurable effect on the magnetic needle. We must regard it as a singular fact that no observations yet made give us the slightest indication as to what this emanation is. The possibility of defining it is suggested by the discovery within the past few years, that under certain conditions, heated matter sends forth entities known as Rontgen rays, Becquerel corpuscles and electrons. I cannot speak authoritatively on this subject, but, so far as I am aware, no direct evidence has yet been gathered showing that any of these entities reach us from the sun. We must regard the search for the unknown agency so fully proved as among the most important tasks of the astronomical physicist of the present time. From what we know of the history of scientific discovery, it seems highly probable that, in the course of his search, he will, before he finds the object he is aiming at, discover many other things of equal or greater importance of which he had, at the outset, no conception.

The main point I desire to bring out in this review is the tendency which it shows towards unification in physical research. Heretofore differentiation— the subdivision of workers into a continually increasing number of groups of specialists—has been the rule. Now we see a coming together of what, at first sight, seem the most widely separated spheres of activity. What two branches could be more widely separated than that of stellar statistics, embracing the whole universe within its scope, and the study of these newly discovered emanations, the product of our laboratories, which seem to show the existence of corpuscles smaller than the atoms of matter? And yet, the phenomena which we have reviewed, especially the relation of terrestrial magnetism to the solar activity, and the formation of nebulous masses around the new stars, can be accounted for only by emanations or forms of force, having probably some similarity with the corpuscles, electrons, and rays which we are now producing in our laboratories. The nineteenth century, in passing away, points with pride to what it has done. It has become a word to symbolize what is most important in human progress Yet, perhaps its greatest glory may prove to be that the last thing it did was to lay a foundation for the physical science of the twentieth century. What shall be discovered in the new fields is, at present, as far without our ken as were the modern developments of electricity without the ken of the investigators of one hundred years ago. We cannot guarantee any special discovery. What lies before us is an illimitable field, the existence of which was scarcely suspected ten years ago, the exploration of which may well absorb the activities of our physical laboratories, and of the great mass of our astronomical observers and investigators for as many generations as were required to bring electrical science to its present state. We of the older generation cannot hope to see more than the beginning of this development, and can only tender our best wishes and most hearty congratulations to the younger school whose function it will be to explore the limitless field now before it.

THE RELATION OF SCIENTIFIC METHOD TO SOCIAL PROGRESS

[Footnote: An address before the Washington Philosophical Society]
Among those subjects which are not always correctly apprehended, even by educated men, we may place that of the true significance of scientific method and the relations of such method to practical affairs. This is especially apt to be the case in a country like our own, where the points of contact between the scientific world on the one hand, and the industrial and political world on the other, are fewer than in other civilized countries. The form which this misapprehension usually takes is that of a failure to appreciate the character of scientific method, and especially its analogy to the methods of practical life. In the judgment of the ordinary intelligent man there is a wide distinction between theoretical and practical science. The latter he considers as that science directly applicable to the building of railroads, the construction of engines, the invention of new machinery, the construction of maps, and other useful objects. The former he considers analogous to those philosophic speculations in which men have indulged in all ages without leading to any result which he considers practical. That our knowledge of nature is increased by its prosecution is a fact of which he is quite conscious, but he considers it as terminating with a mere increase of knowledge, and not as having in its method anything which a person devoted to material interests can be expected to appreciate.

This view is strengthened by the spirit with which he sees scientific investigation prosecuted. It is well understood on all sides that when such investigations are pursued in a spirit really recognized as scientific, no merely utilitarian object is had in view. Indeed, it is easy to see how the very fact of pursuing such an object would detract from that thoroughness of examination which is the first condition of a real advance. True science demands in its every research a completeness far beyond what is apparently necessary for its practical applications. The precision with which the astronomer seeks to measure the heavens and the chemist to determine the relations of the ultimate molecules of matter has no limit, except that set by the imperfections of the instruments of research. There is no such division recognized as that of useful and useless knowledge. The ultimate aim is nothing less than that of bringing all the phenomena of nature under laws as exact as those which govern the planetary motions.

Now the pursuit of any high object in this spirit commands from men of wide views that respect which is felt towards all exertion having in view more elevated objects than the pursuit of gain. Accordingly, it is very natural to classify scientists and philosophers with the men who in all ages have sought after learning instead of utility. But there is another aspect of the question which will show the relations of scientific advance to the practical affairs of life in a different light. I make bold to say that the greatest want of the day, from a purely practical point of view, is the more general introduction of the scientific method and the scientific spirit into the discussion of those political and social problems which we encounter on our road to a higher plane of public well being. Far from using methods too refined for practical purposes, what most distinguishes scientific from other thought is the introduction of the methods of practical life into the discussion of abstract general problems. A single instance will illustrate the lesson I wish to enforce.

The question of the tariff is, from a practical point of view, one of the most important with which our legislators will have to deal during the next few years. The widest diversity of opinion exists as to the best policy to be pursued in collecting a revenue from imports. Opposing interests contend against one another without any common basis of fact or principle on which a conclusion can be reached. The opinions of intelligent men differ almost as widely as those of the men who are immediately interested. But all will admit that public action in this direction should be dictated by one guiding principle—that the greatest good of the community is to be sought after. That policy is the best which will most promote this good. Nor is there any serious difference of opinion as to the nature of the good to be had in view; it is in a word the increase of the national wealth and prosperity. The question on which opinions fundamentally differ is that of the effects of a higher or lower rate of duty upon the interests of the public. If it were possible to foresee, with an approach to certainty, what effect a given tariff would have upon the producers and consumers of an article taxed, and, indirectly, upon each member of the community in any way interested in the article, we should then have an exact datum which we do not now possess for reaching a conclusion. If some superhuman authority, speaking with the voice of infallibility, could give us this information, it is evident that a great national want would be supplied. No question in practical life is more important than this: How can this desirable knowledge of the economic effects of a tariff be obtained?

The answer to this question is clear and simple. The subject must be studied in the same spirit, and, to a certain extent, by the same methods which have been so successful in advancing our knowledge of nature. Every one knows that, within the last two centuries, a method of studying the course of nature has been introduced which has been so successful in enabling us to trace the sequence of cause and effect as almost to revolutionize society. The very fact that scientific method has been so successful here leads to the belief that it might be equally successful in other departments of inquiry.

The same remarks will apply to the questions connected with banking and currency; the standard of value; and, indeed, all subjects which have a financial bearing. On every such question we see wide differences of opinion without any common basis to rest upon.

It may be said, in reply, that in these cases there are really no grounds for forming an opinion, and that the contests which arise over them are merely those between conflicting interests. But this claim is not at all consonant with the form which we see the discussion assume. Nearly every one has a decided opinion on these several subjects; whereas, if there were no data for forming an opinion, it would be unreasonable to maintain any whatever. Indeed, it is evident that there must be truth somewhere, and the only question that can be open is that of the mode of discovering it. No man imbued with a scientific spirit can claim that such truth is beyond the power of the human intellect. He may doubt his own ability to grasp it, but cannot doubt that by pursuing the proper method and adopting the best means the problem can be solved. It is, in fact, difficult to show why some exact results could not be as certainly reached in economic questions as in those of physical science. It is true that if we pursue the inquiry far enough we shall find more complex conditions to encounter, because the future course of demand and supply enters as an uncertain element. But a remarkable fact to be considered is that the difference of opinion to which we allude does not depend upon different estimates of the future, but upon different views of the most elementary and general principles of the subject. It is as if men were not agreed whether air were elastic or whether the earth turns on its axis. Why is it that while in all subjects of physical science we find a general agreement through a wide range of subjects, and doubt commences only where certainty is not attained, yet when we turn to economic subjects we do not find the beginning of an agreement?

No two answers can be given. It is because the two classes of subjects are investigated by different instruments and in a different spirit. The physicist has an exact nomenclature; uses methods of research well adapted to the objects he has in view; pursues his investigations without being attacked by those who wish for different results; and, above all, pursues them only for the purpose of discovering the truth. In economic questions the case is entirely different. Only in rare cases are they studied without at least the suspicion that the student has a preconceived theory to support. If results are attained which oppose any powerful interest, this interest can hire a competing investigator to bring out a different result. So far as the public can see, one man's result is as good as another's, and thus the object is as far off as ever. We may be sure that until there is an intelligent and rational public, able to distinguish between the speculations of the charlatan and the researches of the investigator, the present state of things will continue. What we want is so wide a diffusion of scientific ideas that there shall be a class of men engaged in studying economic problems for their own sake, and an intelligent public able to judge what they are doing. There must be an improvement in the objects at which they aim in education, and it is now worth while to inquire what that improvement is.

It is not mere instruction in any branch of technical science that is wanted. No knowledge of chemistry, physics, or biology, however extensive, can give the learner much aid in forming a correct opinion of such a question as that of the currency. If we should claim that political economy ought to be more extensively studied, we would be met by the question, which of several conflicting systems shall we teach? What is wanted is not to teach this system or that, but to give such a training that the student shall be able to decide for himself which system is right.

It seems to me that the true educational want is ignored both by those who advocate a classical and those who advocate a scientific education. What is really wanted is to train the intellectual powers, and the question ought to be, what is the best method of doing this? Perhaps it might be found that both of the conflicting methods could be improved upon. The really distinctive features, which we should desire to see introduced, are two in number: the one the scientific spirit; the other the scientific discipline. Although many details may be classified under each of these heads, yet there is one of pre-eminent importance on which we should insist.

The one feature of the scientific spirit which outweighs all others in importance is the love of knowledge for its own sake. If by our system of education we can inculcate this sentiment we shall do what is, from a public point of view, worth more than any amount of technical knowledge, because we shall lay the foundation of all knowledge. So long as men study only what they think is going to be useful their knowledge will be partial and insufficient. I think it is to the constant inculcation of this fact by experience, rather than to any reasoning, that is due the continued appreciation of a liberal education. Every business-man knows that a business-college training is of very little account in enabling one to fight the battle of life, and that college-bred men have a great advantage even in fields where mere education is a secondary matter. We are accustomed to seeing ridicule thrown upon the questions sometimes asked of candidates for the civil service because the questions refer to subjects of which a knowledge is not essential. The reply to all criticisms of this kind is that there is no one quality which more certainly assures a man's usefulness to society than the propensity to acquire useless knowledge. Most of our citizens take a wide interest in public affairs, else our form of government would be a failure. But it is desirable that their study of public measures should be more critical and take a wider range. It is especially desirable that the conclusions to which they are led should be unaffected by partisan sympathies. The more strongly the love of mere truth is inculcated in their nature the better this end will be attained.

The scientific discipline to which I ask mainly to call your attention consists in training the scholar to the scientific use of language. Although whole volumes may be written on the logic of science there is one general feature of its method which is of fundamental significance. It is that every term which it uses and every proposition which it enunciates has a precise meaning which can be made evident by proper definitions. This general principle of scientific language is much more easily inculcated by example than subject to exact description; but I shall ask leave to add one to several attempts I have made to define it. If I should say that when a statement is made in the language of science the speaker knows what he means, and the hearer either knows it or can be made to know it by proper definitions, and that this community of understanding is frequently not reached in other departments of thought, I might be understood as casting a slur on whole departments of inquiry. Without intending any such slur, I may still say that language and statements are worthy of the name scientific as they approach this standard; and, moreover, that a great deal is said and written which does not fulfil the requirement. The fact that words lose their meaning when removed from the connections in which that meaning has been acquired and put to higher uses, is one which, I think, is rarely recognized. There is nothing in the history of philosophical inquiry more curious than the frequency of interminable disputes on subjects where no agreement can be reached because the opposing parties do not use words in the same sense. That the history of science is not free from this reproach is shown by the fact of the long dispute whether the force of a moving body was proportional to the simple velocity or to its square. Neither of the parties to the dispute thought it worth while to define what they meant by the word "force," and it was at length found that if a definition was agreed upon the seeming difference of opinion would vanish. Perhaps the most striking feature of the case, and one peculiar to a scientific dispute, was that the opposing parties did not differ in their solution of a single mechanical problem. I say this is curious, because the very fact of their agreeing upon every concrete question which could have been presented ought to have made it clear that some fallacy was lacking in the discussion as to the measure of force. The good effect of a scientific spirit is shown by the fact that this discussion is almost unique in the history of science during the past two centuries, and that scientific men themselves were able to see the fallacy involved, and thus to bring the matter to a conclusion.

If we now turn to the discussion of philosophers, we shall find at least one yet more striking example of the same kind. The question of the freedom of the human will has, I believe, raged for centuries. It cannot yet be said that any conclusion has been reached. Indeed, I have heard it admitted by men of high intellectual attainments that the question was insoluble. Now a curious feature of this dispute is that none of the combatants, at least on the affirmative side, have made any serious attempt to define what should be meant by the phrase freedom of the will, except by using such terms as require definition equally with the word freedom itself. It can, I conceive, be made quite clear that the assertion, "The will is free," is one without meaning, until we analyze more fully the different meanings to be attached to the word free. Now this word has a perfectly well-defined signification in every-day life. We say that anything is free when it is not subject to external constraint. We also know exactly what we mean when we say that a man is free to do a certain act. We mean that if he chooses to do it there is no external constraint acting to prevent him. In all cases a relation of two things is implied in the word, some active agent or power, and the presence or absence of another constraining agent. Now, when we inquire whether the will itself is free, irrespective of external constraints, the word free no longer has a meaning, because one of the elements implied in it is ignored.

To inquire whether the will itself is free is like inquiring whether fire itself is consumed by the burning, or whether clothing is itself clad. It is not, therefore, at all surprising that both parties have been able to dispute without end, but it is a most astonishing phenomenon of the human intellect that the dispute should go on generation after generation without the parties finding out whether there was really any difference of opinion between them on the subject. I venture to say that if there is any such difference, neither party has ever analyzed the meaning of the words used sufficiently far to show it. The daily experience of every man, from his cradle to his grave, shows that human acts are as much the subject of external causal influences as are the phenomena of nature. To dispute this would be little short of the ludicrous. All that the opponents of freedom, as a class, have ever claimed is the assertion of a causal connection between the acts of the will and influences independent of the will. True, propositions of this sort can be expressed in a variety of ways connoting an endless number of more or less objectionable ideas, but this is the substance of the matter.

To suppose that the advocates on the other side meant to take issue on this proposition would be to assume that they did not know what they were saying. The conclusion forced upon us is that though men spend their whole lives in the study of the most elevated department of human thought it does not guard them against the danger of using words without meaning. It would be a mark of ignorance, rather than of penetration, to hastily denounce propositions on subjects we are not well acquainted with because we do not understand their meaning. I do not mean to intimate that philosophy itself is subject to this reproach. When we see a philosophical proposition couched in terms we do not understand, the most modest and charitable view is to assume that this arises from our lack of knowledge. Nothing is easier than for the ignorant to ridicule the propositions of the learned. And yet, with every reserve, I cannot but feel that the disputes to which I have alluded prove the necessity of bringing scientific precision of language into the whole domain of thought. If the discussion had been confined to a few, and other philosophers had analyzed the subject, and showed the fictitious character of the discussion, or had pointed out where opinions really might differ, there would be nothing derogatory to philosophers. But the most suggestive circumstance is that although a large proportion of the philosophic writers in recent times have devoted more or less attention to the subject, few, or none, have made even this modest contribution. I speak with some little confidence on this subject, because several years ago I wrote to one of the most acute thinkers of the country, asking if he could find in philosophic literature any terms or definitions expressive of the three different senses in which not only the word freedom, but nearly all words implying freedom were used. His search was in vain.

Nothing of this sort occurs in the practical affairs of life. All terms used in business, however general or abstract, have that well-defined meaning which is the first requisite of the scientific language. Now one important lesson which I wish to inculcate is that the language of science in this respect corresponds to that of business; in that each and every term that is employed has a meaning as well defined as the subject of discussion can admit of. It will be an instructive exercise to inquire what this peculiarity of scientific and business language is. It can be shown that a certain requirement should be fulfilled by all language intended for the discovery of truth, which is fulfilled only by the two classes of language which I have described. It is one of the most common errors of discourse to assume that any common expression which we may use always conveys an idea, no matter what the subject of discourse. The true state of the case can, perhaps, best be seen by beginning at the foundation of things and examining under what conditions language can really convey ideas.

Suppose thrown among us a person of well-developed intellect, but unacquainted with a single language or word that we use. It is absolutely useless to talk to him, because nothing that we say conveys any meaning to his mind. We can supply him no dictionary, because by hypothesis he knows no language to which we have access. How shall we proceed to communicate our ideas to him? Clearly there is but one possible way— namely, through his senses. Outside of this means of bringing him in contact with us we can have no communication with him. We, therefore, begin by showing him sensible objects, and letting him understand that certain words which we use correspond to those objects. After he has thus acquired a small vocabulary, we make him understand that other terms refer to relations between objects which he can perceive by his senses. Next he learns, by induction, that there are terms which apply not to special objects, but to whole classes of objects. Continuing the same process, he learns that there are certain attributes of objects made known by the manner in which they affect his senses, to which abstract terms are applied. Having learned all this, we can teach him new words by combining words without exhibiting objects already known. Using these words we can proceed yet further, building up, as it were, a complete language. But there is one limit at every step. Every term which we make known to him must depend ultimately upon terms the meaning of which he has learned from their connection with special objects of sense.

To communicate to him a knowledge of words expressive of mental states it is necessary to assume that his own mind is subject to these states as well as our own, and that we can in some way indicate them by our acts. That the former hypothesis is sufficiently well established can be made evident so long as a consistency of different words and ideas is maintained. If no such consistency of meaning on his part were evident, it might indicate that the operations of his mind were so different from ours that no such communication of ideas was possible. Uncertainty in this respect must arise as soon as we go beyond those mental states which communicate themselves to the senses of others.

We now see that in order to communicate to our foreigner a knowledge of language, we must follow rules similar to those necessary for the stability of a building. The foundation of the building must be well laid upon objects knowable by his five senses. Of course the mind, as well as the external object, may be a factor in determining the ideas which the words are intended to express; but this does not in any manner invalidate the conditions which we impose. Whatever theory we may adopt of the relative part played by the knowing subject, and the external object in the acquirement of knowledge, it remains none the less true that no knowledge of the meaning of a word can be acquired except through the senses, and that the meaning is, therefore, limited by the senses. If we transgress the rule of founding each meaning upon meanings below it, and having the whole ultimately resting upon a sensuous foundation, we at once branch off into sound without sense. We may teach him the use of an extended vocabulary, to the terms of which he may apply ideas of his own, more or less vague, but there will be no way of deciding that he attaches the same meaning to these terms that we do.

What we have shown true of an intelligent foreigner is necessarily true of the growing child. We come into the world without a knowledge of the meaning of words, and can acquire such knowledge only by a process which we have found applicable to the intelligent foreigner. But to confine ourselves within these limits in the use of language requires a course of severe mental discipline. The transgression of the rule will naturally seem to the undisciplined mind a mark of intellectual vigor rather than the reverse. In our system of education every temptation is held out to the learner to transgress the rule by the fluent use of language to which it is doubtful if he himself attaches clear notions, and which he can never be certain suggests to his hearer the ideas which he desires to convey. Indeed, we not infrequently see, even among practical educators, expressions of positive antipathy to scientific precision of language so obviously opposed to good sense that they can be attributed only to a failure to comprehend the meaning of the language which they criticise.

Perhaps the most injurious effect in this direction arises from the natural tendency of the mind, when not subject to a scientific discipline, to think of words expressing sensible objects and their relations as connoting certain supersensuous attributes. This is frequently seen in the repugnance of the metaphysical mind to receive a scientific statement about a matter of fact simply as a matter of fact. This repugnance does not generally arise in respect to the every-day matters of life. When we say that the earth is round we state a truth which every one is willing to receive as final. If without denying that the earth was round, one should criticise the statement on the ground that it was not necessarily round but might be of some other form, we should simply smile at this use of language. But when we take a more general statement and assert that the laws of nature are inexorable, and that all phenomena, so far as we can show, occur in obedience to their requirements, we are met with a sort of criticism with which all of us are familiar, but which I am unable adequately to describe. No one denies that as a matter of fact, and as far as his experience extends, these laws do appear to be inexorable. I have never heard of any one professing, during the present generation, to describe a natural phenomenon, with the avowed belief that it was not a product of natural law; yet we constantly hear the scientific view criticised on the ground that events MAY occur without being subject to natural law. The word "may," in this connection, is one to which we can attach no meaning expressive of a sensuous relation.

The analogous conflict between the scientific use of language and the use made by some philosophers is found in connection with the idea of causation. Fundamentally the word cause is used in scientific language in the same sense as in the language of common life. When we discuss with our neighbors the cause of a fit of illness, of a fire, or of cold weather, not the slightest ambiguity attaches to the use of the word, because whatever meaning may be given to it is founded only on an accurate analysis of the ideas involved in it from daily use. No philosopher objects to the common meaning of the word, yet we frequently find men of eminence in the intellectual world who will not tolerate the scientific man in using the word in this way. In every explanation which he can give to its use they detect ambiguity. They insist that in any proper use of the term the idea of power must be connoted. But what meaning is here attached to the word power, and how shall we first reduce it to a sensible form, and then apply its meaning to the operations of nature? Whether this can be done, I do not inquire. All I maintain is that if we wish to do it, we must pass without the domain of scientific statement.

Perhaps the greatest advantage in the use of symbolic and other mathematical language in scientific investigation is that it cannot possibly be made to connote anything except what the speaker means. It adheres to the subject matter of discourse with a tenacity which no criticism can overcome. In consequence, whenever a science is reduced to a mathematical form its conclusions are no longer the subject of philosophical attack. To secure the same desirable quality in all other scientific language it is necessary to give it, so far as possible, the same simplicity of signification which attaches to mathematical symbols. This is not easy, because we are obliged to use words of ordinary language, and it is impossible to divest them of whatever they may connote to ordinary hearers.

I have thus sought to make it clear that the language of science corresponds to that of ordinary life, and especially of business life, in confining its meaning to phenomena. An analogous statement may be made of the method and objects of scientific investigation. I think Professor Clifford was very happy in defining science as organized common-sense. The foundation of its widest general creations is laid, not in any artificial theories, but in the natural beliefs and tendencies of the human mind. Its position against those who deny these generalizations is quite analogous to that taken by the Scottish school of philosophy against the scepticism of Hume.

It may be asked, if the methods and language of science correspond to those of practical life, why is not the every-day discipline of that life as good as the discipline of science? The answer is, that the power of transferring the modes of thought of common life to subjects of a higher order of generality is a rare faculty which can be acquired only by scientific discipline. What we want is that in public affairs men shall reason about questions of finance, trade, national wealth, legislation, and administration, with the same consciousness of the practical side that they reason about their own interests. When this habit is once acquired and appreciated, the scientific method will naturally be applied to the study of questions of social policy. When a scientific interest is taken in such questions, their boundaries will be extended beyond the utilities immediately involved, and one important condition of unceasing progress will be complied with.

THE OUTLOOK FOR THE FLYING MACHINE

Mr. Secretary Langley's trial of his flying-machine, which seems to have come to an abortive issue for the time, strikes a sympathetic chord in the constitution of our race. Are we not the lords of creation? Have we not girdled the earth with wires through which we speak to our antipodes? Do we not journey from continent to continent over oceans that no animal can cross, and with a speed of which our ancestors would never have dreamed? Is not all the rest of the animal creation so far inferior to us in every point that the best thing it can do is to become completely subservient to our needs, dying, if need be, that its flesh may become a toothsome dish on our tables? And yet here is an insignificant little bird, from whose mind, if mind it has, all conceptions of natural law are excluded, applying the rules of aerodynamics in an application of mechanical force to an end we have never been able to reach, and this with entire ease and absence of consciousness that it is doing an extraordinary thing. Surely our knowledge of natural laws, and that inventive genius which has enabled us to subordinate all nature to our needs, ought also to enable us to do anything that the bird can do. Therefore we must fly. If we cannot yet do it, it is only because we have not got to the bottom of the subject. Our successors of the not distant future will surely succeed.

This is at first sight a very natural and plausible view of the case. And yet there are a number of circumstances of which we should take account before attempting a confident forecast. Our hope for the future is based on what we have done in the past. But when we draw conclusions from past successes we should not lose sight of the conditions on which success has depended. There is no advantage which has not its attendant drawbacks; no strength which has not its concomitant weakness. Wealth has its trials and health its dangers. We must expect our great superiority to the bird to be associated with conditions which would give it an advantage at some point. A little study will make these conditions clear.

We may look on the bird as a sort of flying-machine complete in itself, of which a brain and nervous system are fundamentally necessary parts. No such machine can navigate the air unless guided by something having life. Apart from this, it could be of little use to us unless it carried human beings on its wings. We thus meet with a difficulty at the first step—we cannot give a brain and nervous system to our machine. These necessary adjuncts must be supplied by a man, who is no part of the machine, but something carried by it. The bird is a complete machine in itself. Our aerial ship must be machine plus man. Now, a man is, I believe, heavier than any bird that flies. The limit which the rarity of the air places upon its power of supporting wings, taken in connection with the combined weight of a man and a machine, make a drawback which we should not too hastily assume our ability to overcome. The example of the bird does not prove that man can fly. The hundred and fifty pounds of dead weight which the manager of the machine must add to it over and above that necessary in the bird may well prove an insurmountable obstacle to success.

I need hardly remark that the advantage possessed by the bird has its attendant drawbacks when we consider other movements than flying. Its wings are simply one pair of its legs, and the human race could not afford to abandon its arms for the most effective wings that nature or art could supply.

Another point to be considered is that the bird operates by the application of a kind of force which is peculiar to the animal creation, and no approach to which has ever been made in any mechanism. This force is that which gives rise to muscular action, of which the necessary condition is the direct action of a nervous system. We cannot have muscles or nerves for our flying-machine. We have to replace them by such crude and clumsy adjuncts as steam-engines and electric batteries. It may certainly seem singular if man is never to discover any combination of substances which, under the influence of some such agency as an electric current, shall expand and contract like a muscle. But, if he is ever to do so, the time is still in the future. We do not see the dawn of the age in which such a result will be brought forth.

Another consideration of a general character may be introduced. As a rule it is the unexpected that happens in invention as well as discovery. There are many problems which have fascinated mankind ever since civilization began which we have made little or no advance in solving. The only satisfaction we can feel in our treatment of the great geometrical problems of antiquity is that we have shown their solution to be impossible. The mathematician of to-day admits that he can neither square the circle, duplicate the cube or trisect the angle. May not our mechanicians, in like manner, be ultimately forced to admit that aerial flight is one of that great class of problems with which man can never cope, and give up all attempts to grapple with it?

The fact is that invention and discovery have, notwithstanding their seemingly wide extent, gone on in rather narrower lines than is commonly supposed. If, a hundred years ago, the most sagacious of mortals had been told that before the nineteenth century closed the face of the earth would be changed, time and space almost annihilated, and communication between continents made more rapid and easy than it was between cities in his time; and if he had been asked to exercise his wildest imagination in depicting what might come—the airship and the flying-machine would probably have had a prominent place in his scheme, but neither the steamship, the railway, the telegraph, nor the telephone would have been there. Probably not a single new agency which he could have imagined would have been one that has come to pass.

It is quite clear to me that success must await progress of a different kind from that which the inventors of flying-machines are aiming at. We want a great discovery, not a great invention. It is an unfortunate fact that we do not always appreciate the distinction between progress in scientific discovery and ingenious application of discovery to the wants of civilization. The name of Marconi is familiar to every ear; the names of Maxwell and Herz, who made the discoveries which rendered wireless telegraphy possible, are rarely recalled. Modern progress is the result of two factors: Discoveries of the laws of nature and of actions or possibilities in nature, and the application of such discoveries to practical purposes. The first is the work of the scientific investigator, the second that of the inventor.

In view of the scientific discoveries of the past ten years, which, after bringing about results that would have seemed chimerical if predicted, leading on to the extraction of a substance which seems to set the laws and limits of nature at defiance by radiating a flood of heat, even when cooled to the lowest point that science can reach—a substance, a few specks of which contain power enough to start a railway train, and embody perpetual motion itself, almost—he would be a bold prophet who would set any limit to possible discoveries in the realm of nature. We are binding the universe together by agencies which pass from sun to planet and from star to star. We are determined to find out all we can about the mysterious ethereal medium supposed to fill all space, and which conveys light and heat from one heavenly body to another, but which yet evades all direct investigation. We are peering into the law of gravitation itself with the full hope of discovering something in its origin which may enable us to evade its action. From time to time philosophers fancy the road open to success, yet nothing that can be practically called success has yet been reached or even approached. When it is reached, when we are able to state exactly why matter gravitates, then will arise the question how this hitherto unchangeable force may be controlled and regulated. With this question answered the problem of the interaction between ether and matter may be solved. That interaction goes on between ethers and molecules is shown by the radiation of heat by all bodies. When the molecules are combined into a mass, this interaction ceases, so that the lightest objects fly through the ether without resistance. Why is this? Why does ether act on the molecule and not the mass? When we can produce the latter, and when the mutual action can be controlled, then may gravitation be overcome and then may men build, not merely airships, but ships which shall fly above the air, and transport their passengers from continent to continent with the speed of the celestial motions.

The first question suggested to the reader by these considerations is whether any such result is possible; whether it is within the power of man to discover the nature of luminiferous ether and the cause of gravitation. To this the profoundest philosopher can only answer, "I do not know." Quite possibly the gates at which he is beating are, in the very nature of things, incapable of being opened. It may be that the mind of man is incapable of grasping the secrets within them. The question has even occurred to me whether, if a being of such supernatural power as to understand the operations going on in a molecule of matter or in a current of electricity as we understand the operations of a steam-engine should essay to explain them to us, he would meet with any more success than we should in explaining to a fish the engines of a ship which so rudely invades its domain. As was remarked by William K. Clifford, perhaps the clearest spirit that has ever studied such problems, it is possible that the laws of geometry for spaces infinitely small may be so different from those of larger spaces that we must necessarily be unable to conceive them.

Still, considering mere possibilities, it is not impossible that the twentieth century may be destined to make known natural forces which will enable us to fly from continent to continent with a speed far exceeding that of the bird.

But when we inquire whether aerial flight is possible in the present state of our knowledge, whether, with such materials as we possess, a combination of steel, cloth, and wire can be made which, moved by the power of electricity or steam, shall form a successful flying-machine, the outlook may be altogether different. To judge it sanely, let us bear in mind the difficulties which are encountered in any flying-machine. The basic principle on which any such machine must be constructed is that of the aeroplane. This, by itself, would be the simplest of all flyers, and therefore the best if it could be put into operation. The principle involved may be readily comprehended by the accompanying figure. A M is the section of a flat plane surface, say a thin sheet of metal or a cloth supported by wires. It moves through the air, the latter being represented by the horizontal rows of dots. The direction of the motion is that of the horizontal line A P. The aeroplane has a slight inclination measured by the proportion between the perpendicular M P and the length A P. We may raise the edge M up or lower it at pleasure. Now the interesting point, and that on which the hopes of inventors are based, is that if we give the plane any given inclination, even one so small that the perpendicular M P is only two or three per cent of the length A M, we can also calculate a certain speed of motion through the air which, if given to the plane, will enable it to bear any required weight. A plane ten feet square, for example, would not need any great inclination, nor would it require a speed higher than a few hundred feet a second to bear a man. What is of yet more importance, the higher the speed the less the inclination required, and, if we leave out of consideration the friction of the air and the resistance arising from any object which the machine may carry, the less the horse-power expended in driving the plane.

[Illustration]

Maxim exemplified this by experiment several years ago. He found that, with a small inclination, he could readily give his aeroplane, when it slid forward upon ways, such a speed that it would rise from the ways of itself. The whole problem of the successful flying-machine is, therefore, that of arranging an aeroplane that shall move through the air with the requisite speed.

The practical difficulties in the way of realizing the movement of such an object are obvious. The aeroplane must have its propellers. These must be driven by an engine with a source of power. Weight is an essential quality of every engine. The propellers must be made of metal, which has its weakness, and which is liable to give way when its speed attains a certain limit. And, granting complete success, imagine the proud possessor of the aeroplane darting through the air at a speed of several hundred feet per second! It is the speed alone that sustains him. How is he ever going to stop? Once he slackens his speed, down he begins to fall. He may, indeed, increase the inclination of his aeroplane. Then he increases the resistance to the sustaining force. Once he stops he falls a dead mass. How shall he reach the ground without destroying his delicate machinery? I do not think the most imaginative inventor has yet even put upon paper a demonstratively successful way of meeting this difficulty. The only ray of hope is afforded by the bird. The latter does succeed in stopping and reaching the ground safely after its flight. But we have already mentioned the great advantages which the bird possesses in the power of applying force to its wings, which, in its case, form the aeroplanes. But we have already seen that there is no mechanical combination, and no way of applying force, which will give to the aeroplanes the flexibility and rapidity of movement belonging to the wings of a bird. With all the improvements that the genius of man has made in the steamship, the greatest and best ever constructed is liable now and then to meet with accident. When this happens she simply floats on the water until the damage is repaired, or help reaches her. Unless we are to suppose for the flying-machine, in addition to everything else, an immunity from accident which no human experience leads us to believe possible, it would be liable to derangements of machinery, any one of which would be necessarily fatal. If an engine were necessary not only to propel a ship, but also to make her float—if, on the occasion of any accident she immediately went to the bottom with all on board—there would not, at the present day, be any such thing as steam navigation. That this difficulty is insurmountable would seem to be a very fair deduction, not only from the failure of all attempts to surmount it, but from the fact that Maxim has never, so far as we are aware, followed up his seemingly successful experiment.

There is, indeed, a way of attacking it which may, at first sight, seem plausible. In order that the aeroplane may have its full sustaining power, there is no need that its motion be continuously forward. A nearly horizontal surface, swinging around in a circle, on a vertical axis, like the wings of a windmill moving horizontally, will fulfil all the conditions. In fact, we have a machine on this simple principle in the familiar toy which, set rapidly whirling, rises in the air. Why more attempts have not been made to apply this system, with two sets of sails whirling in opposite directions, I do not know. Were there any possibility of making a flying-machine, it would seem that we should look in this direction.

The difficulties which I have pointed out are only preliminary ones, patent on the surface. A more fundamental one still, which the writer feels may prove insurmountable, is based on a law of nature which we are bound to accept. It is that when we increase the size of any flying-machine without changing its model we increase the weight in proportion to the cube of the linear dimensions, while the effective supporting power of the air increases only as the square of those dimensions. To illustrate the principle let us make two flying-machines exactly alike, only make one on double the scale of the other in all its dimensions. We all know that the volume and therefore the weight of two similar bodies are proportional to the cubes of their dimensions. The cube of two is eight. Hence the large machine will have eight times the weight of the other. But surfaces are as the squares of the dimensions. The square of two is four. The heavier machine will therefore expose only four times the wing surface to the air, and so will have a distinct disadvantage in the ratio of efficiency to weight.

Mechanical principles show that the steam pressures which the engines would bear would be the same, and that the larger engine, though it would have more than four times the horse-power of the other, would have less than eight times. The larger of the two machines would therefore be at a disadvantage, which could be overcome only by reducing the thickness of its parts, especially of its wings, to that of the other machine. Then we should lose in strength. It follows that the smaller the machine the greater its advantage, and the smallest possible flying-machine will be the first one to be successful.

We see the principle of the cube exemplified in the animal kingdom. The agile flea, the nimble ant, the swift-footed greyhound, and the unwieldy elephant form a series of which the next term would be an animal tottering under its own weight, if able to stand or move at all. The kingdom of flying animals shows a similar gradation. The most numerous fliers are little insects, and the rising series stops with the condor, which, though having much less weight than a man, is said to fly with difficulty when gorged with food.

Now, suppose that an inventor succeeds, as well he may, in making a machine which would go into a watch-case, yet complete in all its parts, able to fly around the room. It may carry a button, but nothing heavier. Elated by his success, he makes one on the same model twice as large in every dimension. The parts of the first, which are one inch in length, he increases to two inches. Every part is twice as long, twice as broad, and twice as thick. The result is that his machine is eight times as heavy as before. But the sustaining surface is only four times as great. As compared with the smaller machine, its ratio of effectiveness is reduced to one-half. It may carry two or three buttons, but will not carry over four, because the total weight, machine plus buttons, can only be quadrupled, and if he more than quadruples the weight of the machine, he must less than quadruple that of the load. How many such enlargements must he make before his machine will cease to sustain itself, before it will fall as an inert mass when we seek to make it fly through the air? Is there any size at which it will be able to support a human being? We may well hesitate before we answer this question in the affirmative.

Dr. Graham Bell, with a cheery optimism very pleasant to contemplate, has pointed out that the law I have just cited may be evaded by not making a larger machine on the same model, but changing the latter in a way tantamount to increasing the number of small machines. This is quite true, and I wish it understood that, in laying down the law I have cited, I limit it to two machines of different sizes on the same model throughout. Quite likely the most effective flying-machine would be one carried by a vast number of little birds. The veracious chronicler who escaped from a cloud of mosquitoes by crawling into an immense metal pot and then amused himself by clinching the antennae of the insects which bored through the pot until, to his horror, they became so numerous as to fly off with the covering, was more scientific than he supposed. Yes, a sufficient number of humming-birds, if we could combine their forces, would carry an aerial excursion party of human beings through the air. If the watch-maker can make a machine which will fly through the room with a button, then, by combining ten thousand such machines he may be able to carry a man. But how shall the combined forces be applied?

The difficulties I have pointed out apply only to the flying-machine properly so-called, and not to the dirigible balloon or airship. It is of interest to notice that the law is reversed in the case of a body which is not supported by the resistance of a fluid in which it is immersed, but floats in it, the ship or balloon, for example. When we double the linear dimensions of a steamship in all its parts, we increase not only her weight but her floating power, her carrying capacity, and her engine capacity eightfold. But the resistance which she meets with when passing through the water at a given speed is only multiplied four times. Hence, the larger we build the steamship the more economical the application of the power necessary to drive it at a given speed. It is this law which has brought the great increase in the size of ocean steamers in recent times. The proportionately diminishing resistance which, in the flying-machine, represents the floating power is, in the ship, something to be overcome. Thus there is a complete reversal of the law in its practical application to the two cases.

The balloon is in the same class with the ship. Practical difficulties aside, the larger it is built the more effective it will be, and the more advantageous will be the ratio of the power which is necessary to drive it to the resistance to be overcome.

If, therefore, we are ever to have aerial navigation with our present knowledge of natural capabilities, it is to the airship floating in the air, rather than the flying-machine resting on the air, to which we are to look. In the light of the law which I have laid down, the subject, while not at all promising, seems worthy of more attention than it has received. It is not at all unlikely that if a skilful and experienced naval constructor, aided by an able corps of assistants, should design an airship of a diameter of not less than two hundred feet, and a length at least four or five times as great, constructed, possibly, of a textile substance impervious to gas and borne by a light framework, but, more likely, of exceedingly thin plates of steel carried by a frame fitted to secure the greatest combination of strength and lightness, he might find the result to be, ideally at least, a ship which would be driven through the air by a steam-engine with a velocity far exceeding that of the fleetest Atlantic liner. Then would come the practical problem of realizing the ship by overcoming the mechanical difficulties involved in the construction of such a huge and light framework. I would not be at all surprised if the result of the exact calculation necessary to determine the question should lead to an affirmative conclusion, but I am quite unable to judge whether steel could be rolled into parts of the size and form required in the mechanism.

In judging of the possibility of commercial success the cheapness of modern transportation is an element in the case that should not be overlooked. I believe the principal part of the resistance which a limited express train meets is the resistance of the air. This would be as great for an airship as for a train. An important fraction of the cost of transporting goods from Chicago to London is that of getting them into vehicles, whether cars or ships, and getting them out again. The cost of sending a pair of shoes from a shop in New York to the residence of the wearer is, if I mistake not, much greater than the mere cost of transporting them across the Atlantic. Even if a dirigible balloon should cross the Atlantic, it does not follow that it could compete with the steamship in carrying passengers and freight.

I may, in conclusion, caution the reader on one point. I should be very sorry if my suggestion of the advantage of the huge airship leads to the subject being taken up by any other than skilful engineers or constructors, able to grapple with all problems relating to the strength and resistance of materials. As a single example of what is to be avoided I may mention the project, which sometimes has been mooted, of making a balloon by pumping the air from a very thin, hollow receptacle. Such a project is as futile as can well be imagined; no known substance would begin to resist the necessary pressure. Our aerial ship must be filled with some substance lighter than air. Whether heated air would answer the purpose, or whether we should have to use a gas, is a question for the designer.

To return to our main theme, all should admit that if any hope for the flying-machine can be entertained, it must be based more on general faith in what mankind is going to do than upon either reasoning or experience. We have solved the problem of talking between two widely separated cities, and of telegraphing from continent to continent and island to island under all the oceans—therefore we shall solve the problem of flying. But, as I have already intimated, there is another great fact of progress which should limit this hope. As an almost universal rule we have never solved a problem at which our predecessors have worked in vain, unless through the discovery of some agency of which they have had no conception. The demonstration that no possible combination of known substances, known forms of machinery, and known forms of force can be united in a practicable machine by which men shall fly long distances through the air, seems to the writer as complete as it is possible for the demonstration of any physical fact to be. But let us discover a substance a hundred times as strong as steel, and with that some form of force hitherto unsuspected which will enable us to utilize this strength, or let us discover some way of reversing the law of gravitation so that matter may be repelled by the earth instead of attracted—then we may have a flying-machine. But we have every reason to believe that mere ingenious contrivances with our present means and forms of force will be as vain in the future as they have been in the past.